YEAR-BOOK

OF THE

ROYAL SOCIETY OF LONDON

1916

Elibron Classics
www.elibron.com

Elibron Classics series.

ISBN 0-543-92540-4 (paperback)
ISBN 0-543-92539-0 (hardcover)

This Elibron Classics Replica Edition is an unabridged facsimile of the edition published in 1916 by Harrison and Sons, London.

YEAR-BOOK

OF THE

ROYAL SOCIETY OF LONDON.

1916.

LONDON:
HARRISON AND SONS, ST. MARTIN'S LANE, W.C.
Printers in Ordinary to His Majesty.
1916.

Price Five Shillings.

No. 20.

HARRISON AND SONS,
PRINTERS IN ORDINARY TO HIS MAJESTY,
ST. MARTIN'S LANE.

CONTENTS.

MEMORANDUM AS TO THE WISHES OF THE COUNCIL IN RESPECT OF BENEFACTIONS TO THE SOCIETY.

From time to time since its foundation, the Royal Society has, through the generosity of benefactors, received funds, now amounting to a very considerable sum.

In the majority of cases the terms of gift have limited the application of the money to certain definite purposes, and, in particular, to the award of medals or other prizes for scientific discoveries or other contributions to the advancement of Natural Knowledge.

Every year the Council have to award several medals, including the Copley, Royal, Rumford, Davy, Darwin, Buchanan, Sylvester, and Hughes Medals, or some of these, and have been led by experience to the conclusion that it is neither to the advantage of the Society nor in the interests of the advancement of Natural Knowledge that this already long list of medals should in future be added to, and that, therefore, no further bequests to be awarded as prizes for past achievements should be accepted by the Society.

They desire, however, to make known that the funds belonging absolutely to the Society, funds tied down by no special directions as to their applications, funds which the Society are free to use for general purposes, are very few indeed. And the President and Council have again and again had the experience that the usefulness of the Society for the advancement of Natural Knowledge has been greatly hampered by the lack of funds of which they could freely make use according to their own judgment.

The President and Council are confident that it would not be difficult, wherever desirable, to associate in some conspicuous manner with any gift to the Society the name of the benefactor, and indeed they would wish to do so.

The President and Council accordingly desire to make it generally known that, while they will willingly receive gifts to be applied to special objects or for the benefit of particular sciences indicated by the donors, they consider that, in view of the varying necessities of Science, the most useful benefactions are those which are given to the Society in general terms for the advancement of Natural Knowledge.

YEAR-BOOK

OF

THE ROYAL SOCIETY.

1916.

CALENDAR.

1916.

JANUARY	1.	*Last day for receiving applications for Government Grants.*
,,	27.	*Ordinary Meeting at* 4.30 P.M.
FEBRUARY	3.	,, ,, ,,
,,	10.	,, ,, ,,
,,	17.	,, ,, ,,
,,	24.	,, ,, ,,
MARCH	2.	,, ,, ,,
,,	9.	,, ,, ,,
,,	16.	,, ,, ,,
,,	23.	,, ,, ,,
,,	25.	*Annual Contributions due.*
,,	30.	*Ordinary Meeting at* 4.30 P.M.
APRIL	6.	,, ,, ,,
,,	13.	,, ,, ,,
MAY	11.	*Election of Fellows at* 4 P.M.
,,	11.	*Ordinary Meeting at* 4.30 P.M.
,,	18.	,, ,, ,,
,,	25.	,, ,, ,,
JUNE	1.	,, ,, ,,
,,	8.	,, ,, ,,
,,	22.	,, ,, ,,
,,	29.	,, ,, ,,
NOVEMBER	2.	,, ,, ,,
,,	9.	,, ,, ,,
,,	16.	,, ,, ,,
,,	23.	,, ,, ,,
,,	30.	*Anniversary Meeting at* 4 P.M. (*Thursday*).
DECEMBER	7.	*Ordinary Meeting at* 4.30 P.M.
,,	14.	,, ,, ,,
,,	31.	*Last day for receiving Certificates of Candidates for Election.*

THE LIST OF THE ROYAL SOCIETY, 1916.

His Sacred Majesty KING GEORGE V., Patron.

Date of Election

1906 HIS ROYAL HIGHNESS THE DUKE OF CONNAUGHT AND STRATHEARN, K.G.

1914 HIS ROYAL HIGHNESS PRINCE ARTHUR OF CONNAUGHT, K.G.

THE COUNCIL.

SIR JOSEPH J. THOMSON, O.M., M.A., D.Sc., LL.D.—President.

SIR ALFRED BRAY KEMPE, M.A., D.C.L.—Treasurer.

PROF. ARTHUR SCHUSTER, Sc.D., Ph.D.—Secretary.

WILLIAM BATE HARDY, M.A.—Secretary.

DUKINFIELD HENRY SCOTT, M.A., Ph.D., LL.D.—Foreign Secretary.

PROF. JOHN GEORGE ADAMI, M.D.

SIR THOMAS CLIFFORD ALLBUTT, K.C.B.—Vice-President.

FREDERICK FROST BLACKMAN, D.Sc.

DUGALD CLERK, D.Sc.

SIR WILLIAM CROOKES, O.M., D.Sc.—Vice-President.

PROF. ARTHUR DENDY, D.Sc.

HENRY JOHN HORSTMAN FENTON, Sc.D.

JOHN STANLEY GARDINER, M.A.

HENRY HEAD, M.D.

GEORGE WILLIAM LAMPLUGH.

PROF. AUGUSTUS EDWARD HOUGH LOVE, D.Sc.

MAJOR PERCY ALEXANDER MACMAHON, D.Sc.

PROF. ARTHUR SMITHELLS, B.Sc.

PROF. ERNEST HENRY STARLING, M.D.

RICHARD THRELFALL, M.A.

SIR PHILIP WATTS, K.C.B.—Vice-President.

This Council will continue till November 30, 1916.

ROBERT HARRISON—Asst. Secretary and Librarian.

OFFICE STAFF.

T. E. James (*Clerk*). A. H. White (*Assistant Librarian*).

F. A. Towle (*Clerk to the Government Grant Committee*). A. I. Mackie.

H. W. Robinson (*Library Assistant*). P. D. Rogers.

Postal Address: Burlington House, Piccadilly, W.

Telephone No.: 905 Mayfair. *Telegraphic Address:* Royal Society, London.

FELLOWS OF THE ROYAL SOCIETY, 1916.

Date of Election		Service on Council, &c.
1876	Abney, Sir William de Wiveleslie, K.C.B. D.C.L. (Dunelm.) D.Sc. (Vict. and Dub.) F.C.S. F.R.A.S., F. Phys. Soc. Adviser in Science to the Board of Education; lately Principal Assistant Secretary, Board of Education, South Kensington; retired Capt. R.E. **Medal: Rumford.** *Measham Hall, Leicestershire; Rathmore Lodge, Bolton Gardens South, South Kensington,* S.W.; *and Athenæum Club.*	1883–85 91–93
1905	Adami, John George, M.A. M.D. (Camb. McGill) F.R.S.E. F.R.S. (Canada) Sc.D. (Trin. Coll. Dublin) LL.D. (Toronto and Univ. New Bruns.) Professor of Pathology, McGill University, Montreal. 34 *Macgregor Street, Montreal, Canada.*	1915–
1907	Adams, Frank Dawson, D.Sc. (McGill), Ph.D. (Heidelb.), Logan Professor of Geology and Dean of the Faculty of Applied Science in McGill University. *McGill University, Montreal.*	
1889	Aitken, John, F.R.S.E. LL.D. (Glasg.). *Ardenlea, Falkirk,* N.B.	
1901	Alcock, Alfred William, Lieut.-Colonel I.M.S. (retired), C.I.E. M.B. LL.D. C.M.Z.S.; some time Superintendent of the Indian Museum, and Professor of Zoology in the Medical College, Calcutta. Hon. Mem. Nederlandsche Dierkund. Vereen., Calif. Acad. Sci., Hon. Fellow Asiat. Soc. Bengal. *Heathlands, Erith Road, Belvedere, Kent.*	1911–13
1880	Allbutt, Sir Thomas Clifford—**Vice-President**—K.C.B. M.A. M.D. LL.D. Hon. D.Sc. (Oxon) F.L.S. Regius Professor of Physic in the University of Cambridge. *St. Radegund's, Cambridge.*	1896–98 1914– **V.P.** 1914–
1914	Allen, Edgar Johnson, D.Sc. (Lond.) F.Z.S. Secretary of the Marine Biological Association of the United Kingdom and Director of the Plymouth Laboratory. *The Laboratory, Citadel Hill, Plymouth.*	
1891	Allerton, Right Hon. William Lawies Jackson, Lord. 41 *Cadogan Square,* S.W.; *and Allerton Hall, Chapel Allerton, Leeds.*	
1907	Anderson, Hugh Kerr, M.A. M.D. *The Lodge, Gonville and Caius College, Cambridge.*	
1915	Andrewes, Frederick William, D.M. M.A. (Oxon) F.R.C.P. Pathologist to St. Bartholomew's Hospital; Professor of Pathology in the University of London. 1 *North Grove, Highgate,* N.	
1906	Andrews, Charles William, D.Sc. (Lond.) F.G.S. *British Museum (Natural History), Cromwell Road,* S.W.	

Date of Election		Service on Council, &c.
1876	Armstrong, Henry Edward, Ph.D. (Lips.) LL.D. (St. Andr.) F.C.S. Emeritus Professor of Chemistry in the Imperial College of Science and Technology, City and Guilds (Engineering) College; Hon. Mem. Pharm. Soc. Lond. **Medal: Davy.** 55 *Granville Park, Lewisham*, S.E.; *and Athenæum Club.*	1888-90 1900-2 **V.P.** 1901-2
1912	Arnold, John Oliver, D. Met. (Sheffield) Professor of Metallurgy in the University of Sheffield. *The University, St. George's Square, Sheffield.*	
1908	Asquith, Right Hon. Herbert Henry, D.C.L. (Oxon) LL.D. (Camb.). 10 *Downing Street*, S.W.	
1902	Baker, Herbert Brereton, M.A. D.Sc. Professor of General Chemistry in the Imperial College of Science and Technology. *Imperial College of Science, South Kensington; and Latchmoor House, Gerrard's Cross, Bucks.*	1909-10
1898	Baker, Henry Frederick, Sc.D. Lowndean Professor of Astronomy and Geometry in the University of Cambridge, Fellow of St. John's College, Mem. Lond. Math. Soc., Fell. Camb. Phil. Soc. **Medal: Sylvester.** *Walcott, Storey's Way, Cambridge.*	1906-08
1878	Baker, John Gilbert, F.L.S. M.R.I.A.; lately Keeper of the Herbarium, Royal Gardens, Kew. 3 *Cumberland Road, Kew.*	1883-84
1888	Balfour, Right Hon. Arthur James, LL.D. D.C.L. Chancellor of the University of Edinburgh. *Whittingehame, Prestonkirk.*	1907-08 1912-14 **V.P.** 1912-14
1884	Balfour, Isaac Bayley, D.Sc. M.D. (Edin.) LL.D. (Glasg.) M.A. (Oxon) F.R.S.E. F.L.S. King's Botanist in Scotland, Regius Keeper of the Royal Botanic Garden, Edinburgh, Professor of Botany in the University of Edinburgh; Hon. Mem. Pharm. Soc. Gt. Britain, Camb. Phil. Soc.; Corresp. Mem. Deutsch. Bot. Gesell., Soc. Nat. Sci. Nat. et Math. Cherbourg, New York Acad. Sci.; Mem. Assoc. Soc. Roy de Bot. de Belg. *Inverleith House, Edinburgh; and Athenæum Club.*	1892-94
1909	Baly, Edward Charles Cyril, M.Sc. (Liverp.) F.I.C. Professor of Inorganic Chemistry in the University of Liverpool; Formerly Assistant Professor of Chemistry and Lecturer on Spectroscopy in the University of London. 14 *Sunnyside, Princes Park, Liverpool.*	
1910	Barcroft, Joseph, M.A. (Camb.). Fellow of King's College and Lecturer in Physiology in the University of Cambridge. *Gleneagh, Grange Road, Cambridge.*	
1912	Barkla, Charles Glover, M.A. (Camb.) D.Sc. (Liverp.) M.Sc. (Vict.) Professor of Natural Philosophy in the University of Edinburgh. *Littledene, Priestfield Road, Edinburgh.*	

Date of Election		Service on Council, &c.
1909	Barlow, Sir Thomas, Bart. K.C.V.O. M.D. (Lond.) Hon. D.Sc. (Vict.) Hon. LL.D. (Aberd.). Physician to His Majesty's Household. 10 *Wimpole Street*, W.	
1908	Barlow, William, F.G.S. F.C.S. *The Red House, Stanmore, Middlesex.*	
1909	Barnes, Rev. Ernest William, M.A. Sc.D. F.R.A.S. F.C.P.S. Lately Fellow and Tutor of Trinity College, Cambridge. Master of the Temple. *The Master's House, Temple*, E.C.	
1911	Barnes, Howard Turner, M.A. M.Sc. D.Sc. F.R.S., Canada. Macdonald Professor of Physics and Director of the Physics Building in McGill University. *McGill University, Montreal, Canada.*	
1899	Barrett, Sir William Fletcher, Kt., F.R.S.E. M.R.I.A. M.I.E.E. Lately Professor of Experimental Physics in the Royal College of Science for Ireland. 31 *Devonshire Place*, W.	
	Barry (*see* Wolfe Barry).	
1889	Basset, Alfred Barnard, M.A. *Fledborough Hall, Holyport, Berks.*	
1894	Bateson, William, M.A., Hon. Fellow of St. John's College, Cambridge, and formerly Professor of Biology in the University. Director of the John Innes Horticultural Institution, Merton, Surrey; Fullerian Professor of Physiology, Royal Institution. **Medal: Darwin.** *The Manor House, Merton, Surrey.*	1901-03
1909	Bather, Francis Arthur, M.A. D.Sc. (Oxon) F.G.S. F.Z.S. Hon. Mem. Soc. Linn. Normandie, Roy. Geol. Soc., Cornwall. Assistant Keeper of the Department of Geology, British Museum (Nat. Hist.). *Fabo, Marryat Road, Wimbledon*, S.W.	
1903	Bayliss, William Maddock, M.A. D.Sc. (Oxon). University Professor of General Physiology in University College, London. **Medal: Royal.** *St. Cuthbert's, West Heath Road, Hampstead*, N.W.	1913-15
1892	Beddard, Frank Evers, M.A. (Oxon) D.Sc. F.R.S.E. F.Z.S. Hon. Mem. Ac. Sci. et Lettres, Christiania. Corresp. Mitgl. d. Kön. Böhm. Ges. d. Wissensch., Hon. Mem. N.Z. Inst. 20 *Sherriff Road, West Hampstead*, N.W.	
1908	Bedford, Herbrand Arthur Russell, Duke of, K.G., Pres. Zool. Soc. *Woburn Abbey, Bedfordshire; and* 15 *Belgrave Square*, S.W.	
1906	Beilby, George Thomas, LL.D. (Glasg.) F.I.C. F.C.S. Assoc. Inst. C.E. 11 *University Gardens, Glasgow.*	
1897	Bell, Robert, I.S.O. M.D. D.Sc. LL.D. Formerly Chief Geologist of the Geological Survey of Canada. *Ottawa, Canada.*	
1907	Benham, William Blaxland, M.A. (Oxon) D.Sc. (Lond.). Professor of Biology in the University of Otago, New Zealand. Formerly Aldrichian Demonstrator in Comparative Anatomy, University of Oxford. *Dunedin, New Zealand.*	
1914	Bennett, Geoffrey Thomas, M.A. Fellow and Mathematical Lecturer of Emmanuel College, Cambridge. *Emmanuel College, Cambridge.*	

Date of Election		Service on Council, &c.
1908	Berkeley, Right Hon. Randal Thomas Mowbray, Earl of, F.C.S. *Foxcombe, near Oxford.*	
1871	Besant, William Henry, Sc.D. F.R.A.S. F.C.P.S. Fellow of St. John's College, Cambridge. *Spring Lawn, Harvey Road, Cambridge.*	
1914	Biffen, Rowland Harry, M.A. Professor of Agricultural Botany in the University of Cambridge. *School of Agriculture, Cambridge.*	
1906	Blackman, Frederick Frost, D.Sc. (Lond.) Fellow of St. John's College, Cambridge, and Reader in Botany in the University. *St. John's College, Cambridge.*	1914–
1913	Blackman, Vernon Herbert, M.A. Sc.D. (Camb.). Professor of Plant Physiology and Pathology in the Imperial College of Science and Technology; formerly Fellow of St. John's College, Cambridge. *Imperial College of Science and Technology, South Kensington,* S.W.; *and* 17 *Berkeley Place, Wimbledon,* S.W.	
1905	Bone, William Arthur, D.Sc. (Vict. Manch.) Ph.D. (Heidelb.), F.C.S. Some time Bishop Berkeley Fellow of Owens College and Fellow of Victoria University, Manchester; Hon. Mem. Inst. Gas. Eng. and Amer. Gas Inst.; Professor of Chemical Technology (Fuel and Refractory Materials) in the Imperial College of Science and Technology. *Montrose, Harpenden Road, St. Albans; and Imperial College of Science and Technology, South Kensington,* S.W.	
1878	Bonney, Rev. Thomas George, Sc.D. (Camb.) LL.D. (McGill) D.Sc. (Dubl., Sheffield) F.S.A. F.G.S. Soc. Phil. Cantab.; Acad. Reg. Hib. et Ebor. Soc. Honor.; Soc. Géol. Belg. et Soc. Reg. Canard. Corresp.; Corresp. Mem. Soc. Géol. du Nord de France; Hon. Canon of Manchester; Emeritus Professor of Geology in University College, London. 9 *Scroope Terrace, Cambridge.*	1880–82 1895 97–99 V.P. 1898–99
1899	Booth, Right Hon. Charles, Hon. Sc.D. (Camb.). 28 *Campden House Court, Kensington,* W.	
1888	Bottomley, James Thomson, M.A. D.Sc. F.R.S.E. F.C.S. 13 *University Gardens, Glasgow.*	
1894	Boulenger, George Albert, LL.D. (St. And.) D.Sc. (Louvain) F.Z.S. Assoc. Acad. Roy. Belg., For. Mem. R. Swedish Acad., Sci., Corresp. Mem. R. Accad. d. Sci., Turin, Acad. Sci., Philadelphia New York, Indiana, Imp. Soc. Friends of Sci., Moscow, Soc. Linn. Bordeaux, Soc. Nat. Acclim. France, Soc. R. Zool.-Malac. Belgique, R. Soc. Espan. Hist. Nat., Madrid, Sci. Soc. Boston, Nat. Hist. Soc. Basel, Mus. Nacion. Para, Hon. Mem. Soc. Sci. Chili, Soc. Natur., Luxemburg, Soc. Cent. d'Aquicult. de France. 8 *Courtfield Road South Kensington; and British Museum (Nat. Hist.), Cromwell Road* S.W.	1903–05

Date of Election		Service on Council, &c.
1895	Bourne, Sir Alfred Gibbs, K.C.I.E., D.Sc. Lately Professor of Biology in the Presidency College, Madras. Fellow of University College, London. *Middlepark, Paignton, S. Devon.*	
1910	Bourne, Gilbert Charles, M.A. D.Sc. (Oxon) F.L.S. F.Z.S. Fellow of Merton College and Linacre Professor of Comparative Anatomy in the University of Oxford; formerly Fellow and Tutor of New College, Oxford. *Savile House, Oxford.*	
1891	Bower, Frederick Orpen, Sc.D. (Camb.) F.L.S. F.R.S.E. Regius Professor of Botany in the University of Glasgow; Hon. Mem. Deutsch. Botan. Gesell.; Corr. Mem. K. B. Akad. Wissen. Munich. **Medal: Royal.** 1 *St. John's Terrace, Hillhead, Glasgow.*	1901–02
1914	Boycott, Arthur Edwin, M.A. D.M. Professor of Pathology, Victoria University, Manchester. 17 *Loom Lane, Radlett.*	
1888	Boys, Charles Vernon, A.R.S.M. Officier de l'Instr. Publ. France, Hon. Mem. New York Acad. Sci., Phys. Soc., Moscow. **Medal: Royal.** 66 *Victoria Street*, S.W.; *and Athenæum Club.*	1900–02
1894	Bradford, Sir John Rose, K.C.M.G. C.B. M.D. D.Sc. F.R.C.P. F.Z.S. Hon. M.D. Christiania; Physician to University College Hospital; formerly Professor of Medicine in University College, and Professor Superintendent, Brown Institution, London. 8 *Manchester Square*, W.; *and Athenæum Club.*	1903–04 1907–15 **For. Sec.** 1907–08 **Sec.** 1908–15
1882	Brady, George Stewardson, M.D. LL.D. D.Sc. Hon. Professor of Natural History in the Armstrong College of Durham University, Newcastle. *Park Hurst, Endcliffe, Sheffield.*	
1907	Bragg, William Henry, M.A. (Camb.). D.Sc. (Manch., Brown Univ. Rhode Island). Quain Professor of Physics in the University of London; Lately Cavendish Professor of Physics in the University of Leeds; Nobel Laureate, Physics, 1915. Barnard medallist, Colombia University, 1914. *University of London, University College, Gower Street*, W.C.	1911–12
1904	Brodie, Thomas Gregor, M.D. Professor of Physiology in the University of Toronto; formerly Professor-Superintendent, Brown Institution, University of London, and Professor of Physiology, Royal Veterinary College. *University, Toronto, Canada.*	
1906	Bromwich, Thomas John I'Anson, Sc.D. Fellow and Lecturer of St. John's College, Cambridge, and University Lecturer in Mathematics; Hon. Sec. Lond. Math. Soc.; formerly Professor of Mathematics in Queen's College, Galway, and Fellow of the Royal University of Ireland. 1 *Selwyn Gardens, Cambridge.*	
1911	Brown, Adrian John, M.Sc. (Birm.) F.C.S. F.I.C. Professor of Brewing, and Biology and Chemistry of Fermentation, in the University of Birmingham. *West Heath House, Northfield, Birmingham.*	
1879	Brown, Alexander Crum, D.Sc. LL.D. M.D. Lately Professor of Chemistry in the University of Edinburgh. 8 *Belgrave Crescent, Edinburgh.*	1891–92
1898	Brown, Ernest William, M.A. Sc.D. Professor of Mathematics, Yale University, New Haven. **Medal: Royal.** 116 *Everit Street, New Haven, Conn., U.S.A.*	

Date of Election		Service on Council, &c.
1889	Brown, Horace T., LL.D. (Edin.) F.C.S. F.I.C. F.G.S. F.L.S. **Medal: Royal.** 52 *Nevern Square, Kensington*, S.W.	1899–1901
1883	Browne, Sir James Crichton, Kt., M.D. LL.D. D.Sc. (Leeds) F.R.S.E. 45 *Hans Place*, S.W.; *and Athenæum Club.*	
1899	Bruce, Sir David, Surgeon-General, R.A.M.S., Kt., C.B. M.B. C.M. F.R.C.P. Hon. D.Sc. (Dublin) LL.D. (Glasgow). Director Royal Society Commissions, Uganda 1903, Malta 1904–6, Uganda 1908–10, Nyasaland 1911–13. Corresp. Étranger, Acad. de Méd., Paris; Mem. Hon. de la Soc. de Pathologie Exotique, Paris; Socio ororario dell' Accad. di Scienze, Lettere, ed Arti, Acireale, Sicily. **Medal: Royal.** *Royal Army Medical College, Grosvenor Road*, S.W.; *and Athenæum Club.*	1904–05
1874	Brunton, Sir T. Lauder, Bart., M.D. Sc.D. LL.D. (Edin.) Hon. LL.D. (Aberdeen) Coll. Reg. Med. Soc. 1 *De Walden Court, New Cavendish Street*, W.; *and Athenæum Club.*	1882–84 1905–06 V.P. 1905–06
1895	Bryan, George Hartley, Sc.D. Professor of Mathematics in the University College of North Wales. *Plas Gwyn, Bangor, N. Wales.*	
1893	Bryce, Right Hon. James, Viscount, O.M. D.C.L. Lately His Majesty's Ambassador at Washington, U.S.A.; Hon. Fellow, Trinity and Oriel Colleges, Oxford; For. Mem. Inst. de France; Pruss. Ord. "*Pour le Mérite*"; Mem. Acad. Roy. des Sci. Brux., R. Accad. delle Sci. Torino, Soc. Romana di Storia Patria, Reale Accad. dei Lincei, Rome; Academies of Petrograd, Stockholm, and Naples. Massachusetts Hist. Soc. *Hindleap, Forest Row, Sussex.*	1899–1900
1887	Buchanan, John Young, M.A. F.R.S.E. F.C.S. F.R.G.S. Comm. de l'Ordre de St. Charles de Monaco; Vice-prés. du Comité de perfectionnement de l'Inst. Océanographique (Fondation Albert Ier. Prince de Monaco); Hon. Mem. Gesell. für Erdkunde, Berlin, Schw. Naturf. Gesell. 26 *Norfolk Street, Park Lane*, W.	
1913	Bulloch, William, M.D. Professor of Bacteriology in the University of London. *London Hospital Medical College, Turner Street, Mile End Road*, E.	
1893	Burnside, William, M.A. D.Sc. (Dubl.) LL.D. (Edin.) Hon. Fellow of Pembroke College, Cambridge; Professor of Mathematics, Royal Naval College, Greenwich. **Medal: Royal.** *The Croft, Bromley Road, Catford*, S.E.	1901–02
1904	Burrard, Sir Sidney Gerald, Colonel R.E., K.C.S.I. Surveyor-General of India. Formerly Superintendent of the Trigonometrical Survey of India. *Dehra Dun, India.*	
1894	Callendar, Hugh Longbourne, M.A.; Professor of Physics in the Imperial College of Science and Technology, London; lately Fellow of Trinity College, Cambridge; LL.D. (McGill Univ.) F.R.S. (Canada). **Medal: Rumford.** 49 *Grange Road, Ealing*, W.	1902–04

Date of Election		Service on Council, &c.
1905	Campbell, John Edward, M.A. (Oxon). Fellow and Tutor of Hertford College, Oxford, and Mathematical Lecturer in University College, Oxford. 14 *Rawlinson Road, Oxford.*	
1871	Carruthers, William, F.L.S. F.G.S. F.R.P.S.E. Ph.D. (Upsala) Fell. Bot. Soc. Edin.; Corr. Mem. Acad. Nat. Sci. Philad., New York Acad. Sci., Ist. Ven. Sci. Lett. ed Art., Soc. Bot. Copenh., Soc. Géol. Belg.; Hon. Mem. Manch. Lit. and Phil. Soc., Whitby Lit. and Phil. Soc., Chester Nat. Hist. Soc., Nat. Hist. Soc. Glasg., Dumf. and Gall. Nat. Hist. and Antiq. Soc.; F.R.M.S.; F.L.S.; lately Keeper Botanical Department, British Museum; Consulting Botanist, Royal Agricultural Society of England. 44 *Central Hill, Norwood,* S.E.	1877–79
1887	Cash, John Theodore, M.D. LL.D. (Edin.) Regius Professor of Materia Medica in the University of Aberdeen. *Marischal College, Aberdeen.*	
1913	Chapman, David Leonard, M.A. (Oxon). Fellow of Jesus College, Oxford. *Jesus College, Oxford.*	
1907	Chattaway, Frederick Daniel, M.A. (Oxon) D.Sc. (Lond.) Ph.D. (Munich). 151 *Woodstock Road, Oxford.*	
1894	Cheyne, Sir William Watson, Bart., K.C.M.G. C.B. M.B. C.M. (Edin.) Pres. R. Coll. Surg.; Professor of Surgery in King's College, London. 75 *Harley Street,* W.	1901–02
1897	Chree, Charles, M.A. Sc.D. (Camb.) LL.D. (Aberd.) Superintendent of Kew Observatory. *Old Deer Park, Richmond, Surrey.*	
1881	Christie, Sir William Henry Mahoney, K.C.B. M.A. Hon. D.Sc. (Oxon). Formerly Astronomer Royal. F.R.A.S., F.R. Met. Soc., Corr. Mem. Acad. Sci. Paris, and Imp. Acad. Sci. Petrograd; For. Memb. Roy. Acad. Sci. Palermo; Corr. Mem. Soc. Spettros. Ital., and Soc. Nationale des Sci. Nat. et Math. Cherbourg. *The Tower House, Down, Kent.*	1883–85 89–91 1900–01 V.P. 1890–91 1900–01
1872	Cleland, John, M.D. D.Sc. LL.D. Formerly Professor of Anatomy in the University of Glasgow. *Drumclog, Crewkerne, Somerset.*	1897–98
1908	Clerk, Dugald, D.Sc. (Manch.) M.Inst.C.E. M.Inst.M.E. F.C.S. 57 and 58 *Lincoln's Inn Fields,* W.C.; *Lukyns, Ewhurst, Surrey; and Athenæum Club.*	1915–
1868	Clifton, Robert Bellamy, M.A. (Cantab. et Oxon) F.R.A.S. Lately Professor of Experimental Philosophy in the University of Oxford; Soc. Lit. Phil. Manc. Soc. Honor. 3 *Bardwell Road, Oxford; and Athenæum Club.*	1871–73 85–87 96–98 V.P. 1896–98
1912	Cockayne, Leonard, Ph.D. (Munich.) F.L.S. Corr. Mem. Bot. Soc. Edin., Pres. Phil. Inst. Cant. N.Z. 20 *Colombo Street, Wellington, New Zealand.*	

Date of Election		Service on Council, &c.
1911	Cohen, Julius Berend, B.Sc. (Vict.) Ph.D. (Munich). Professor of Organic Chemistry in the University of Leeds. 1 *North Grange Mount, Headingley, Leeds.*	
1910	Coleman, Arthur Philemon, M.A. Ph.D. Professor of Geology, University of Toronto. *Geological Department, University, Toronto; and* 476 *Huron Street, Toronto, Canada.*	
1896	Collie, J. Norman, Ph.D. LL.D. F.R.S.E. F.R.G.S. F.I.C. F.C.S. Professor of Organic Chemistry in the University of London. 16 *Campden Grove, Kensington,* W.	1905–07
1915	Conway, Arthur William, M.A. (Oxon, R. Univ. Ireland), Hon. D.Sc. (R. Univ. Ireland), M.R.I.A., *Elsinore, Coliemore Road, Dalkey, Co. Dublin.*	
1903	Copeman, Sydney Monckton, M.A. M.D. (Camb.) F.R.C.P. D.P.H. F.Z.S. Medical Inspector to H.M. Local Government Board; and Lecturer on Hygiene and Public Health, Westminster Hospital Medical School. **Medal: Buchanan.** *Local Government Board, Whitehall,* S.W.; *and* 76 *Regent's Park Road,* N.W.	
1878	Cotterill, James Henry, M.A. Formerly Professor of Applied Mechanics, Royal Naval College, Greenwich. *Hill Crest, Parkstone, Dorset.*	
1906	Cowell, Philip Herbert, M.A. (Camb.) Hon. D.Sc. (Oxon). Superintendent of the Nautical Almanac. Formerly Chief Assistant in the Royal Observatory, Greenwich. 62 *Shooter's Hill Road, Blackheath,* S.E.	
1885	Creak, Ettrick William, C.B. Captain R.N. F.R.G.S. Mem. Inst. Elect. Eng.; Hon. Memb. Roy. Cornwall Polytech. Soc. 9 *Hervey Road, Blackheath,* S.E.	1898–1900
1911	Cromer, Rt. Hon. Evelyn Baring, Earl of, G.C.B. O.M. G.C.M.G. K.C.S.I. C.I.E. 36 *Wimpole Street,* W.	
1863	Crookes, Sir William, Kt. O.M.—**Past President—Vice-President**—Hon. LL.D. (Birm.) Hon. D.Sc. (Oxon, Sheffield, Cape of Good Hope), Hon. Sc.D. (Camb. Dubl.), F.C.S., Mem. Inst. Elect. Eng., Soc. Psych. Research; Hon. Mem. Roy. Phil. Soc. Glasgow, Roy. Soc. N.S.W., Pharm. Soc., Chem. Metall. and Mining Soc. of South Africa, Amer. Chem. Soc., Amer. Philos. Soc., Roy. Soc. Sci. Upsala, Deutsch. Chem. Gesell. Berlin, Psychol. Soc. Paris, "Antonio Alzate" Sci. Soc. Mexico, Sci. Soc. Bucharest, Reg. Accad. Zelanti, Acireale; Corresp. Inst. France (Acad. Sci.), Corresp. Mem. Bataafsch Genoots. Rotterdam, Soc. d'Encouragement pour l'Indust. Paris; For. Mem. R. Accad. Lincei., Rome. **Medals: Copley, Royal, Davy.** 7 *Kensington Park Gardens,* W.; *and Athenæum Club.*	1877–79 94–96 1907– **For. Sec.** 1908–12 **V.P.** 1895–96 1907–08 1915– **Pres.** 1913–15

Date of Election		Service on Council, &c.
1907	Crossley, Arthur William, D.Sc. (Vict.) Ph.D. (Würzburg). F.C.S., F.I.C.; University Professor of Chemistry in the University of London (King's College), lately Berkeley Fellow of the Owens College; formerly Lecturer in Chemistry at St. Thomas's Hospital; 46 *Lindfield Gardens, Hampstead*, N.W.; *and King's College, Strand*, W.C.	
1898	Curzon of Kedleston, Right Hon. George Nathaniel, Earl, K.G. G.C.S.I. G.C.I.E. D.C.L. D.L. Chancellor of the University of Oxford. F.R.G.S. 1 *Carlton House Terrace*, S.W.	
1907	Cushny, Arthur Robertson, M.A. M.D. LL.D. Professor of Pharmacology and Materia Medica in the University of London, University College. 8 *Upper Park Road, Hampstead*, N.W.	
1914	Cuthbertson, Clive, B.A. (Oxon), Fellow of University College, London University. 58 *Kensington Park Road*, W.	
1913	Dalby, William Ernest, M.A. M.Inst.C.E., M.Inst.Mech.E. Professor of Civil and Mechanical Engineering in the City and Guilds Engineering College, Imperial College of Science and Technology, London. 37 *Madeley Road, Ealing*, W.	
1914	Dale, Henry Hallett, M.A. M.D. (Camb.), Director of the Wellcome Physiological Research Laboratories, Herne Hill. 140 *Thurlow Park Road, Dulwich*, S.E.	
1882	Darwin, Sir Francis, Kt. M.A. Sc.D. and M.B. (Cantab.). LL.D. (St. And.) D.Sc. (Dubl. Liverpool, Sheffield, Brussels) Hon. Ph.D. (Upsala, Prague) F.L.S. F.Z.S., Hon. Fellow of Christ's College; lately Reader in Botany in the Univ. of Cambridge. Hon. Mem. Moscow Univ., Hon. Fellow, Soc. Imp. Nat. Moscow; For. Mem. Soc. Holl. Sci. Haarlem; Mem. Soc. Nat. Sci. et Math. de Cherbourg, K. Leop.-Carol. Acad. Halle, Soc. "Antonio Alzate," Mexico, Amer. Phil. Soc.; Corr. Mem. Imp. Acad. Sci. Petrograd, Deutsch. Botan. Gesell., Sencken. Natur. Gesell. Frankfort, Inst. Nat. Geneva. **Medal: Darwin.** 10 *Madingley Road, Cambridge.*	1894–95 1902–08 **For. Sec.** 1903–07 **V.P.** 1907–08
1903	Darwin, Horace, M.A. (Camb.). *The Orchard, Cambridge.*	
1900	David, T. W. Edgeworth, C.M.G. B.A. (Oxon), Hon. D.Sc. (Oxon), F.G.S. Professor of Geology in the University of Sydney. *The University, Sydney, N.S.W.*	
1867	Dawkins, W. Boyd, M.A. D.Sc. (Oxon, Manch.) F.S.A. F.G.S. M. Inst. M.E. Assoc. Inst. C.E. Hon. Fellow of Jesus Coll., Oxford; Honorary Professor of Geology and Palæontology in the Victoria University of Manchester; Soc. Anthrop. Berol., Acad. Sci. Nat. Philad. et Soc. Nat. Hist. Bost. Corresp. Soc. Phil. Amer. et Acad. Sci. Nov. Ebor. et Soc. Geol. Belg. Soc. Honor. Pres. Cambrian Archæol. Assoc. and Somerset Archæol. and Nat. Hist. Soc. *Fallowfield House, Fallowfield, Manchester; and Athenæum Club.*	1889–91

Date of Election		Service on Council, &c.
1908	Dendy, Arthur, D.Sc. F.L.S. Professor of Zoology in the University of London (King's College); formerly Professor of Biology in the Canterbury College (University of New Zealand), and Professor of Zoology in the South African College (Cape Town); Associate Owens College; Hon. Mem. N.Z. Inst.; Corr. Mem. Roy. Socs. Victoria and Tasmania. *Vale Lodge, Hampstead*, N.W.	1915–
1914	Devonshire, Right Hon. Victor Christian William Cavendish, Duke of, K.G. G.C.V.O. LL.D. Chancellor of Leeds University. 78 *Piccadilly*, W.	
1877	Dewar, Sir James, Kt., M.A. F.C.S. F.I.C. F.R.S.E. Hon. LL.D. (Edin. Aberd. Glasg. and St. And.) D.Sc. (Oxon, Dubl., Univ. Ireland, Vict. Brussels) Hon. Ph.D. (Christiania, Freiburg). For. Mem. Nat. Acad. Sci. U.S.A., R. Accad. dei Lincei, Rome, Deutsch. Chem. Gesell. Berlin, Acad. Roy. Sci. Belg.; Hon. Mem. Inst. C.E., Lit. and Phil. Soc. Manc., Pharm. Soc. Lond., New York Acad. Sci., Phil. Soc. Philad., Amer. Chem. Soc., Phil. Soc. Glasg., Soc. Phys. Verein. Frankfurt, R. Ist. Lomb. di Scienze, Lettere ed Arti, Milan; Corr. Mem. Accad. Sci. Turin; Mem. K. Dansk. Videns. Selskab, Copenhagen, Soc. "Antonio Alzate," Mexico; Fellow of Peterhouse, Cambridge; Jacksonian Prof. of Natural Experimental Philosophy in the University of Cambridge; Fullerian Prof. of Chemistry in the Royal Institution. **Medals: Rumford, Davy.** 1 *Scroope Terrace, Cambridge; and Royal Institution Albemarle Street*, W.	1885–86 1898–1900 **V.P.** 1899–1900
1905	Dines, William Henry, B.A. (Camb.) F.R. Met. Soc. *Benson, Wallingford, Berks.*	
1910	Dixey, Frederick Augustus, M.A. D.M. (Oxon). Fellow, Bursar and Lecturer of Wadham College, Oxford. *Wadham College, Oxford.*	
1904	Dixon, Alfred Cardew, Sc.D. (Camb.) M.A. (Lond.) F.C.P.S. Professor of Mathematics, Queen's University, Belfast. *Hurstwood, Malone Park, Belfast.*	
1912	Dixon, Arthur Lee, M.A. (Oxon), Fellow and Tutor of Merton College. *Merton College, Oxford.*	
1886	Dixon, Harold Baily, M.A. (Oxon). M.Sc. (Manch.) Ph.D. (Prag.) F.C.S. Professor of Chemistry and Director of the Chemical Laboratories in the University of Manchester. **Medal: Royal.** *University, Manchester; and Beechey House, Wilbraham Road, Fallowfield, Manchester.*	1902–04
1908	Dixon, Henry Horatio, ScD. (Dubl.). Professor of Botany in Dublin University; Director of Trinity College Botanic Gardens. *School of Botany, Trinity College, Dublin; and Clevedon, Temple Road, Dublin.*	

Date of Election		Service on Council, &c.
1911	Dixon, Walter Ernest, M.A. (Camb.) M.D. B.S. B.Sc. (Lond.) D.P.H. Hon. Mem. Pharm. Soc.; Professor of Pharmacology in King's College, London, and University Lecturer in Pharmacology, Cambridge. *The Grove, Whittlesford, Cambridge.*	
1904	Dobbie, Sir James Johnston, Kt., M.A. D.Sc. LL.D. Government Chemist, London. Formerly Director of the Royal Scottish Museum, Edinburgh. 4 *Vicarage Gate, Kensington,* W.	
1915	Doncaster, Leonard, Sc.D. (Camb.) F.Z.S. Fellow and Lecturer of King's College, Cambridge; University Lecturer in Zoology. 3 *Millington Road, Cambridge.*	
1911	Donnan, Frederick George, M.A. (Roy. Univ. Ireland) Ph.D. (Leip.). Professor of Inorganic and Physical Chemistry in the University of London. *Chemical Laboratories, University College,* W.C.	
1896	Downing, Arthur Matthew Weld, M.A. D.Sc. (Dubl.) F.R.A.S.; formerly Superintendent of the Nautical Almanac. 12 *Oakcroft Road, Blackheath,* S.E.	
1855	Ducie, Right Hon. Henry John Moreton, Earl of, G.C.V.O. F.G.S. *Tortworth Court, Falfield, Gloucestershire; and* 16 *Portman Square,* W.	
1907	Duddell, William, M.Inst.Elect.Eng. **Medal: Hughes.** 47 *Hans Place,* S.W.	
1893	Dunstan, Wyndham R., C.M.G. M.A. (Oxon.) LL.D. (Aberd.) F.C.S. F.I.C.; Director of the Imperial Institute; Hon. Mem. Aristotelian Soc.; Corr. Mem. Inst. Égyptien; formerly Professor of Chemistry to the Pharmaceutical Society of Great Britain and Lecturer on Chemistry at St. Thomas' Hospital. *Imperial Institute, South Kensington,* S.W.	1905-07
	Dyer (*see* Thiselton-Dyer).	
1901	Dyson, Sir Frank Watson, Kt. M.A. (Camb.) LL.D. (Edin.) D.Sc. (Perth, Melb.) F.R.S.E. F.R.A.S. Astronomer Royal: formerly Astronomer Royal for Scotland; Corr. Mem. Bureau des Longitudes, Corr. Mem. Acad. Sci. Paris. *Royal Observatory, Greenwich; and Athenæum Club.*	1903-05 1912-14 V.P. 1913-14
1914	Eddington, Arthur Stanley, M.A. (Camb.) M.Sc. (Manch.) F.R.A.S. Plumian Professor of Astronomy and Experimental Philosophy in the University of Cambridge; Fellow of Trinity College. *The Observatory, Cambridge.*	
1891	Elliott, Edwin Bailey, M.A. F.R.A.S. Waynflete Professor of Pure Mathematics in the University of Oxford; Fellow of Magdalen College, Oxford. 4 *Bardwell Road, Oxford.*	1899-1901
1913	Elliott, Thomas Renton, M.A. M.D. (Camb.). Assistant Physician, University College Hospital. *University College Hospital, Gower Street,* W.C.	

Date of Election		Service on Council, &c.
1893	Ellis, William, F.R.A.S. F.R. Met. Soc. M.I.E.E.; lately Superintendent of the Magnetical and Meteorological Department, Royal Observatory, Greenwich. 78 *Coleraine Road, Blackheath*, S.E.	
1897	Elwes, Henry John, F.L.S. F.Z.S. F.E.S. *Colesborne Park, near Cheltenham.*	
1869	Esson, William, M.A. F.C.S. F.R.A.S. Savilian Professor of Geometry in the University of Oxford, Fellow of New College, Senior Bursar of Merton College. *Merton College; and* 13 *Bradmore Road, Oxford.*	
1901	Evans, Sir Arthur John, Kt. M.A. D.Litt. (Oxon.) LL.D. (Edin.) D. Litt. (Dubl.) Pres. S.A. Fellow of Brasenose College; Hon. Keeper of the Ashmolean Museum, Extraordinary Professor of Prehistoric Archæology, Oxford. *Youlbury, Oxford.*	
1915	Evershed, John, F.R.A.S. Director, Kodaikanal and Madras Observatories. *The Observatory, Kodaikanal, South India.*	
1893	Ewart, James Cossar, M.D. (Edin.) F.R.S.E. F.Z.S. Regius Professor of Natural History in the University of Edinburgh; late Professor of Natural History in the University of Aberdeen; Hon. Mem. Soc. Cent. d'Aquicult. de France. *Craigybield, Penecuik, Midlothian; and Athenæum Club*, S.W.	1907–09 **V.P.** 1908–09
1887	Ewing, Sir J. Alfred, K.C.B. Hon. M.A. (Camb.) LL.D. (St. And. Edin.) F.R.S.E. M. Inst. C.E.; Director of Naval Education; Hon. Mem. Lit. Phil. Soc. Manchester; Corresp. Reale Accad. Sci. Turin. **Medal: Royal.** *Froghole, Edenbridge, Kent; and Athenæum Club.*	1896–98
1900	Farmer, John Bretland, M.A. D.Sc. (Oxon) F.L.S. Professor of Botany in the Imperial College of Science and Technology, London. *South Park, Gerrard's Cross, Bucks.*	1905–06
1899	Fenton, Henry John Horstman, M.A. Sc.D. F.I.C. F.C.S., Honorary Fellow of Christ's College, Cambridge, University Lecturer in Chemistry, Cambridge. 19 *Brookside, Cambridge.*	1913–15 1915–
1876	Ferrier, Sir David, Kt., M.A. (Aberd.) M.D. (Edin.) LL.D. F.R.C.P. Emeritus Professor of Neuro-pathology, King's College, London; Consulting Physician to King's College Hospital and to the National Hospital for Paralysed and Epileptic. **Medal: Royal.** 34 *Cavendish Square*, W.; *and Athenæum Club.*	1886–88 1906–08 **V.P.** 1906–08
1905	Field, Sir Arthur Mostyn, K.C.B., Admiral. Lately Hydrographer of the Navy. F.R.A.S. F.R.G.S. *Serle's Hill, Twyford, Winchester; and United Service Club.*	
1913	Fields, John Charles, Ph.D. (Johns Hopkins). Professor of Mathematics in the University of Toronto. *Faculty Union, University, Toronto, Canada.*	

Date of Election		Service on Council, &c.
1910	Filon, Louis Napoleon George, M.A. D.Sc. Goldsmid Professor of Applied Mathematics and Mechanics in the University of London. *Lynton, Haling Park Road, Croydon.*	
1892	Fleming, John Ambrose, M.A. (Camb.) D.Sc. (Lond.) lately Fellow of St. John's College, Cambridge; Fellow of University College, London, and Professor of Electrical Engineering in the University of London. **Medal: Hughes.** *Pender Electrical Laboratory, University College, Gower Street,* W.C.	
1889	Fletcher, Sir Lazarus, Kt. M.A. and Hon. Fellow, Univ. Coll. Oxon. Hon. LL.D. (St. And.) Hon. Ph.D. (Berlin) F.G.S. F.C.S. Mem. Mineralog. Soc., Corr. Mem. New York Acad. Sci., K. Gesell. Wissen. Gött., Munich, Soc. Cient. "Antonio Alzate," Mexico. Director of the Natural History Departments, British Museum. *Natural History Museum, Cromwell Road; and* 35 *Woodville Gardens, Ealing,* W.; *and Athenæum Club.*	1895-97 1910-12 **V.P.** 1910-12
1915	Fletcher, Walter Morley, M.A. M.D. Sc.D. Fellow and formerly Senior Tutor of Trinity College, Cambridge. Secretary of the National Medical Research Committee. 71 *Bedford Gardens, Campden Hill,* W.	
1913	Flett, John Smith, M.A. M.B. C.M. D.Sc. (Edin.) LL.D. (Edin.) F.G.S. F.R.S.E. Assistant (for Scotland) to Director of H.M. Geological Survey. *H.M. Geological Survey,* 33 *George Square, Edinburgh; and* 13 *West Castle Road, Edinburgh.*	
1887	Forbes, George, M.A. (Camb.) F.R.S.E. F.R.A.S. Mem. Inst. C.E. M.I.E.E. Chev. Lég. Honor. Mem. Astron. Gesell. Vienna, Amer. Phil. Soc., and Franklin Inst.; formerly Professor of Nat. Phil. in Anderson's College, Glasgow. 11 *Little College Street, Westminster,* S.W.	
1905	Forster, Martin Onslow, D.Sc. (Lond.) Ph.D. (Würzb.) F.I.C. Formerly Assistant Professor of Chemistry in the Imperial College of Science and Technology, London. 84 *Cornwall Gardens,* S.W.	
1886	Forsyth, Andrew Russell, M.A. Sc.D. (Camb.) Hon. D.Sc. (Oxon) Hon. Sc.D. (Dubl. Vict. Liverp., Calcutta) Hon. LL.D. (Glasg. Aberd.) Hon. Math. D. (Christiania) F.C.P.S. F.R.A.S. Hon. F.R.S.E. Mem. Lond. Math. Soc. Hon. Mem. Lit. Phil. Soc. Manch., Soc. Corr. R. Ist. Lomb. Milan, R. Ist. Venice, Nat. Acad. Sci. Washington, Math. Soc. Kharkov, Math. Soc. Calcutta; formerly Sadleirian Professor of Pure Mathematics in the University of Cambridge and Fellow of Trinity College, Cambridge; Chief Professor of Mathematics in the Imperial College of Science and Technology, South Kensington. **Medal: Royal.** 8 *Brompton Square,* S.W.	1893-95
1869	Foster, George Carey, B.A. (Lond.) LL.D. (Glasg.) D.Sc. (Vict. Manch.) F.C.S. F. Phys. Soc. Lond. Mem. Inst. Elect. Eng.; Fellow of the University of London, and of University Coll. London; Hon. Mem. Jewish Histor. Soc., Amer. Phil. Soc. Lately Principal, and formerly Professor of Physics in University College, London. *Ladywalk, Rickmansworth, Herts; and Athenæum Club.*	1870-72 77-78 83-85 91-93 1901-03 **V.P.** 1891-93 1902-03

Date of Election		Service on Council, &c
1910	Fowler, Alfred, A.R.C.S. F.R.A.S. Assistant Professor of Physics (Astrophysics Department), Imperial College of Science and Technology, London. 19 *Rusthall Avenue, Bedford Park, W.*	
1891	Frankland, Percy Faraday, Ph.D. (Würzb.), B.Sc. (Lond.) M.Sc. (Birm.) LL.D. (St. Andr.) Sc.D. (Dubl.) A.R.S.M., F.C.S. F.I.C. Professor of Chemistry in the University of Birmingham, Hon. Mem. Pharm. Soc., Soc. Publ. Anal. *The University, Edgbaston, Birmingham.*	1903-05
1877	Fraser, Sir Thomas Richard, Kt. M.D. (Edin. Dubl.) F.R.C.P. & F.R.S. (Edin.) Sc.D. (Camb.) LL.D. (Aberd. and Glasg.). Professor of Materia Medica and Clinical Medicine in the University of Edinburgh. Hon. Physician to the King in Scotland. 13 *Drumsheugh Gardens, Edinburgh.*	
1894	Froude, Robert Edmund, C.B. Superintendent of the Admiralty Experimental Works, Gosport. *North Lodge, Alverstoke, Gosport.*	
1883	Fry, Right Hon. Sir Edward, G.C.B. B.A. (Lond.) D.C.L. (Oxon) LL.D. (Camb. Aberd. Edin. Bristol) F.S.A. F.L.S. Fellow of the University of London; and Hon. Fellow, Balliol Coll., Oxon. *Failand House, Failand, near Bristol.*	
1892	Gadow, Hans Friedrich, Ph.D. (Jena) Hon. M.A. (Camb.) Strickland Curator and Lecturer on the Advanced Morphology of Vertebrates in the University of Cambridge. *Zoological Laboratory, Cambridge.*	1899-1901
1907	Gamble, Frederick William, D.Sc. (Manch.) M.Sc. (Birm.) F.Z.S. Mason Professor of Zoology and Comparative Anatomy in the University of Birmingham. *The University, Edmund Street, Birmingham.*	
1899	Gamble, James Sykes, C.I.E. M.A. (Oxon) F.L.S., lately Conservator of Forests in India, and Director of the Imperial Forest School, Dehra Dun. *Highfield, East Liss, Hants; and Athenæum Club.*	
1908	Gardiner, John Stanley, M.A. F.L.S. F.R.G.S. F.Z.S. Fellow of Gonville and Caius College, Cambridge. Professor of Zoology and Comparative Anatomy in the University. *Bredon House, Selwyn Gardens, Cambridge; and Zoological Laboratory, Cambridge.*	1911-12 1914–
1890	Gardiner, Walter, Sc.D. (Camb.) M.A. Fellow and Bursar of Clare College, Cambridge. **Medal: Royal.** *St. Awdreys, Hill's Road, Cambridge.*	
1910	Garrod, Archibald Edward, D.M. M.A. (Oxon) LL.D. (Aberd.) F.R.C.P. Physician to St. Bartholomew's Hospital; Consulting Physician to the Hospital for Sick Children. 9 *Chandos Street, Cavendish Square, W.*	1914-15

Date of Election		Service on Council, &c.
1914	Garwood, Edmund Johnston, M.A. Sc.D. Yates Goldsmid Professor of Geology and Mineralogy in the University of London. F.G.S. *The Pryors, Hampstead,* N.W.	
1865	Geikie, Sir Archibald, O.M. K.C.B.—**Past President**—D.C.L. (Oxon) Sc.D. (Cantab. Dubl.) LL.D. (Edin. Glasg. St. And. Aberd. Liverpl. Birmingham) Ph.D. (Leipzig, Upsala, Prague) F.R.S.E. F.G.S., lately Director-General of the Geological Survey of the United Kingdom, and of the Museum of Practical Geology, London, Officier de la Légion d'Honneur; Corresp. Inst. France (Acad. Sci.); Corr. Mem. Acad. Berlin, Munich, Turin, Valdarnese del Poggio, Philad., K. Gesell. Wissen. Gött., Nat. Hist. Soc. Boston, Geograph. Soc. Italy, Netherlands; For. Mem. R. Accad. Lincei, Rome, Acad. Petrograd, Vienna, Stockholm, Christiania, Naples, Amer. Acad. Sci. Boston; Assoc. Nat. Acad. Sci. Washington, Acad. Roy. Sci. Belgium; Hon. Mem. Royal Irish Acad., Inst. Civ. Eng., Phil. Soc. Camb., York, Fed. Inst. Min. Eng., University Moscow, K. Leop.-Carol. Deutsch Akad. Halle, Soc. Españ. Hist. Nat. Madrid, Accad. Sci. Acireale, Imp. Soc. Natur. St. Petersburg, Soc. Natur. Moscow, Soc. Min. St. Petersburg, Geol. Soc. Edinburgh, Glasgow, Liverpool, Manchester, Stockholm, Acad. Sci. New York, National Geograph. Soc. Washington, Soc. Geograph. Ital. **Medal: Royal.** *Shepherd's Down, Haslemere, Surrey.*	1885–87 89–93 1903–13 **Sec.** 1903–08 **For. Sec.** 1889–93 **V.P.** 1885–87 **Pres.** 1908–13
1891	Gilchrist, Percy Carlyle, A.R.S.M. M.Inst.C.E. M.Inst.M.E. M.R.I.; Chevalier Lég. Hon. France. *Reform Club, Pall Mall,* S.W.	
1875	Glaisher, James Whitbread Lee, Sc.D. (Camb. and Dubl.) F.R.A.S. F.C.P.S. Mem. Lond. Math. Soc. **Medal: Sylvester.** *Trinity College, Cambridge.*	1883–84 90–92
1882	Glazebrook, Richard Tetley, C.B. M.A. Hon. Sc.D. (Vict.) F.C.P.S. M.Inst. Elect. Eng.; Fellow of Trinity College, Cambridge; Director of the National Physical Laboratory; formerly Principal of University College, Liverpool. **Medal: Hughes.** *Bushy House, Teddington, Middlesex; and Athenæum Club.*	1892–94
1882	Godman, Frederick Du Cane, D.C.L. (Oxon) F.L.S. F.G.S. F.E.S. a Trustee of the British Museum. 45 *Pont Street,* S.W.; *and South Lodge, Horsham.*	1891–93
1880	Godwin-Austen, Henry Haversham, Lieut.-Col. F.Z.S. F.R.G.S. *Nore, Godalming.*	
1902	Goldie, Right Hon. Sir George D. Taubman, K.C.M.G. F.R.C.S. *Naval and Military Club,* 94 *Piccadilly,* W.	
1905	Goodrich, Edwin Stephen, M.A. (Oxon) F.L.S. Sec. L.S. Aldrichian Demonstrator of Comparative Anatomy in the University of Oxford; Fellow of Merton College. 53 *Banbury Road; and Merton College, Oxford.*	

Date of Election		Service on Council, &c.
1896	Gorst, Right Hon. Sir John Eldon, Kt. K.C. M.A. Hon. Fellow of St. John's College, Cambridge. 84 *Campden Hill Court*, W.	1901–03
1908	Gowland, William, A.R.S.M. F.I.C. F.C.S. F.S.A. Mem. Iron and Steel Inst.; Pres. Inst. of Metals; Mem. Inst. Mining and Metallurgy and Roy. Anthrop. Inst.; Hon. Mem. Roy. Cornwall Polytech. Soc.; Chevalier of the Order of the Rising Sun. Emeritus Professor of Metallurgy in the Royal School of Mines. 13 *Russell Road, Kensington*, W.	1912–14
1908	Grace, John Hilton, M.A. Lecturer in Mathematics at Peterhouse and at Pembroke College, Cambridge. *Peterhouse, Cambridge.*	
1896	Gray, Andrew, M.A. LL.D. (Glasg.) F.R.S.E. Professor of Natural Philosophy in the University of Glasgow. 11 *The University, Glasgow.*	
1915	Green, Arthur George, M.Sc. (Leeds) F.I.C. F.C.S. Professor of Tinctorial Chemistry in the University of Leeds. *University, Leeds; and* 49 *Cardigan Road, Headingley, Leeds.*	
1888	Greenhill, Sir A. George, Kt. M.A. Lately Professor of Mathematics in the Ordnance College, Woolwich; Officier d'Académie, Paris; For. Mem. R. Accad. dei Lincei. **Medal: Royal.** 1 *Staple Inn*, W.C.	1896–97
1878	Greenwell, Rev. William, M.A. D.C.L. F.S.A. Rector of St. Mary-in-the-South Bailey, Durham. *Durham.*	
1901	Gregory, John Walter, D.Sc. F.G.S. Professor of Geology in the University of Glasgow. *The University, Glasgow.*	
1914	Grey, Right Hon. Sir Edward, Bt. K.G. D.C.L. *Foreign Office*, S.W.; *and Fallodon, Christon Bank, R.S.O., Northumberland.*	
1895	Griffiths, Ernest Howard, Sc.D. (Camb.) D.Sc. (Wales and Victoria) LL.D. (Aberd.). Principal of University College of South Wales and Monmouthshire. Formerly Fellow, now Hon. Fellow, of Sidney Sussex College, Cambridge. *University College, Cardiff; and* 24 *Newport Road, Cardiff.* **Medal: Hughes.**	1909–11
1883	Groves, Charles Edward, F.C.S. F.I.C. 352 *Kennington Road*, S.E.	
1883	Grubb, Sir Howard, Kt. F.R.A.S. Hon. M.A.I. (Dubl.), Mem. Roy. Dubl. Soc., Hon. Mem. Roy. Inst. Civ. Eng. Ireland, Astron. Soc. Mexico. *Aberfoyle, Rathgar, Dublin.*	
1899	Haddon, Alfred Cort, M.A. Sc.D. M.R.I.A. Reader in Ethnology, University of Cambridge. 3 *Cranmer Road, Cambridge.*	

Date of Election		Service on Council, &c.
1909	Hadfield, Sir Robert Abbott, Kt. D.Sc. D.Met. M.Inst.C.E. Mem. Iron and Steel Institute. 22 *Carlton House Terrace*, S.W.	
1897	Haldane, John Scott, M.A. (Oxon) M.D. (Edin.) LL.D. (Birm.) Fellow of New College, and lately Reader in Physiology in the University of Oxford. *Cherwell, Oxford.*	1908–09
1906	Haldane, Right Hon. Richard Burdon, Viscount, K.T. O.M. M.A. (Edin.) Hon. D.C.L. (Oxon), Chancellor of the University of Bristol. 28 *Queen Anne's Gate*, S.W.; *and Athenæum Club.*	
1909	Hall, Alfred Daniel, M.A. (Oxon). Commissioner, Development Fund, *The Development Commission*, 6A *Dean's Yard*, S.W.; *and Ewhurst, Mostyn Road, Merton*, S.W.	
1891	Halliburton, William Dobinson, M.D. B.Sc. F.R.C.P. LL.D. (Aberdeen, Toronto), Professor of Physiology in King's College, London. *Church Cottage*, 17 *Marylebone Road*, N.W.	1898–1900 1903–04
1887	Halsbury, Right Hon. Hardinge Stanley Giffard, Earl of, M.A. D.C.L. High Steward of the University of Oxford. 4 *Ennismore Gardens*, W.	
1863	Harcourt, Augustus George Vernon, M.A. Hon. D.Sc. (Oxon) D.C.L. (Dunelm.) LL.D. (Univ. McGill) V.P.C.S. Lately Lee's Reader in Chemistry at Christ Church, Oxford. *St. Clare, Ryde, Isle of Wight; and Athenæum Club.*	1878–80
1909	Harden, Arthur, D.Sc. (Manch.) Ph.D. (Erlangen). Professor of Biochemistry (London University); Head of the Department of Biochemistry, Lister Institute; formerly Lecturer in Chemistry, Victoria University, Manchester. 5 *Cambridge Gardens, Marlborough Road, Richmond, Surrey.*	
1910	Hardy, Godfrey Harold, M.A. Fellow and Lecturer of Trinity College, Cambridge. *Trinity College, Cambridge.*	
1902	Hardy, William Bate—**Secretary**—M.A. Fellow and Tutor of Gonville and Caius College. *Newnham Lea, Grange Road, Cambridge.*	1911–13 1914– **V.P.** 1914–15 **Sec.** 1915–
1902	Harker, Alfred, M.A. LL.D. (McGill). Fellow of St. John's College and Lecturer in Petrology in the University of Cambridge. *St. John's College, Cambridge.*	
1910	Harker, John Allen, D.Sc. (Tüb.) M.Sc. (Manch.). Chief Assistant and Head of the Thermometry Division at the National Physical Laboratory; formerly Dalton Chemical Scholar and Bishop Berkeley Fellow in Physics at the Owens College, Manchester. *Alston, Queen's Road, Teddington*, S.W.	
1898	Harmer, Sidney Frederic, Sc.D. late Fellow of King's College, Cambridge. Keeper of the Department of Zoology, British Museum (Natural History). *Hillymead, Thornton Hill, Wimbledon*, S.W.	1905–06

Date of Election | Service on Council, &c.

1897 Haswell, William, M.A. D.Sc. (Edin.) F.L.S. Corr. Mem. Roy. Soc. Tasman.; Mem. K. Leop. Carol. Deutsche Akad. Halle; Corr. Mem. Soc. Biol. Paris; Challis Professor of Biology in the University of Sydney. *The University, Sydney, N.S.W.*

1914 Havelock, Thomas Henry, M.A. (Camb.) D.Sc. (Durh.). Professor of Applied Mathematics in Armstrong College, Newcastle-on-Tyne. *Rockliffe, Gosforth, Newcastle-on-Tyne.*

1915 Hayden, Henry Hubert, C.I.E. B.A.I. (Dubl.) Hon. D.Sc. (Calcutta) F.G.S. Director, Geological Survey of India. *Geological Survey of India, Calcutta.*

1899 Head, Henry, M.A. M.D. F.R.C.P. M.R.C.S. **Medal: Royal.** 1915– 4 *Montagu Square*, W.

1906 Heape, Walter, Hon. M.A. (Camb.) F.Z.S. 10 *King's Bench Walk, Temple*, E.C.

1912 Heath, Sir Thomas Little, K.C.B. Sc.D. (Camb.) Hon. D.Sc. (Oxon.); formerly Fellow of Trinity College, Cambridge; Joint Permanent Secretary to the Treasury and Auditor of the Civil List. 64 *Bedford Gardens, Kensington*, W.

1891 Heaviside, Oliver, Hon. Ph.D. (Gött.). Hon. Mem. Lit. Phil. Soc. Manchester, Inst. Elect. Eng.; Amer. Acad. Arts and Sci. *Homefield, Lower Warberry, Torquay.*

1899 Hele-Shaw, Henry Selby, D.Sc. (Bristol) LL.D. (St. Andr.) M.Inst.C.E. M.Inst.M.E., Assoc. Inst. Naval Arch., F.R. Met. Soc., formerly Principal of the Transvaal Technical Inst. (afterwards University Coll.); and Organizer of Technical Education in the Transvaal. 64 *Victoria Street, Westminster*, S.W.

1889 Hemsley, William Botting, LL.D. (Aberd.) F.L.S. Hon. Mem. Roy. Soc. N.S.W. N.Z. Inst.; Corr. Mem. Deutsch. Botan. Gesell.; formerly Keeper of the Herbarium, Royal Gardens, Kew. 24 *Southfield Gardens, Strawberry Hill.*

1874 Henrici, Olaus Magnus Friedrich Erdmann, Ph.D. LL.D. (St. And.) 1882–83 Formerly Professor of Mechanics and Mathematics in the Central Technical College of the City and Guilds of London Institute. *Hiltingbury Lodge, Chandler's Ford, Hants.*

1892 Herdman, William Abbott, D.Sc. F.R.S.E. F.L.S. D.Sc. (Harvard); 1898–1900 General Secretary of the British Association; Professor of Natural History in the University of Liverpool. *Croxteth Lodge, Ullet Road, Liverpool; and Rowany, Port Erin, Isle of Man.*

1871 Herschel, John, Col. R.E. F.R.A.S. Lately Deputy Superintendent, Great Trigonometrical Survey of India. *Observatory House, Slough, Bucks.*

1910 Hewitt, John Theodore, M.A. (Camb.) D.Sc. (Lond.) Ph.D. (Heidelb.) A.R.C.S. F.C.S. Professor of Chemistry in the University of London (East London College), Mile End, E. *Clifford House, Staines Road, Bedfont, Middlesex.*

Date of Election		Service on Council, &c.
1895	Heycock, Charles Thomas, M.A. Fellow, and Lecturer in Natural Science, King's College, Cambridge; Goldsmiths' Reader in Metallurgy. 3 *St. Peter's Terrace, Cambridge.*	1907–09
1885	Hicks, William Mitchinson, M.A. D.Sc. Late Fellow of St. John's College, Cambridge; Professor of Physics in the University of Sheffield. **Medal: Royal.** *Leamhurst, Ivy Park Road, Sheffield.*	1900–02 1910–11
1895	Hickson, Sydney John, D.Sc. (Lond.) M.A. (Camb.) Hon. M.A. (Oxon) F.Z.S. Hon. Fellow of Downing College, Cambridge; Professor of Zoology in the University of Manchester; Hon. Mem. K. Inst. Taal-Land-en Volkenkunde Neder.-Indië; Dr. Bot.-Zool., Groningen. *Ellesmere House, Withington, Manchester.*	1906–08
1903	Hiern, William Philip, M.A. (Camb. Oxon.) F.L.S. Corr. Mem. R. Acad. Lisb. *The Castle, Barnstaple, Devon.*	
1913	Hill, James Peter, D.Sc. (Edin.) F.L.S. F.Z.S. Jodrell Professor of Zoology in the University of London. *Kanimbla, Dollis Avenue, Finchley,* N.	
1900	Hill, Leonard Erskine, M.B. Lately Professor of Physiology, University of London. Director of the Department of Applied Physiology, Medical Research Committee. *Osborne House, Loughton, Essex.*	
1894	Hill, Micaiah J. M., M.A. Sc.D. Hon. LL.D. (St. And.) Professor of Mathematics in the University of London. 18 *Ferncroft Avenue, Hampstead,* N.W.	1911–13
1911	Hills, Edmond Herbert, C.M.G. Col. R.E. (retired); formerly Head of the Topographical Department of the War Office. F.R.A.S. 32 *Prince's Gardens,* S.W.	
1896	Hinde, George Jennings, Ph.D. (Munich) F.G.S. 24 *Avondale Road, South Croydon.*	
1913	Hinks, Arthur Robert, M.A. (Camb.) F.R.A.S. Secretary, Royal Geographical Society; Gresham Lecturer in Astronomy. 109 *Corringham Road, Golder's Green,* N.W.	
1893	Hobson, Ernest William, Sc.D. Sadleirian Professor of Pure Mathematics in the University of Cambridge. Fellow of Christ's College, Cambridge; Hon. D.Sc. (Oxon, Dubl. Manch. Sheffield), LL.D. (St. Andr.); For. Mem. K. Leop.-Carol. Akad. Halle. **Medal: Royal.** *The Gables, Mount Pleasant, Cambridge.*	1903–05
1895	Holden, Sir Henry Capel Lofft, K.C.B. Col. R.A. (retired), Mem. Inst. Elect. Eng. 2 *St. John's Park, Blackheath.*	
1904	Holland, Sir Thomas Henry, K.C.I.E. A.R.C.S. D.Sc. F.G.S. M.Inst.M.M. M.Inst.M.E. Professor of Geology and Mineralogy in the University of Manchester. Formerly Berkeley Fellow of the Owens College; formerly Director of the Geological Survey of India; Hon. Mem. Min. and Geol. Inst. of India, and Roy. Cornwall Polyt. Soc.; F. Asiat. Soc., Bengal. *Westwood, Alderley Edge, Cheshire.*	

Date of Election		Service on Council, &c.
1905	Hopkins, Frederick Gowland, M.A. (Camb.) D.Sc. M.B. (Lond.). M.R.C.S. M.R.C.P. F.I.C. Reader in Chemical Physiology in the University of Cambridge; Prælector in Biochemistry and Fellow of Trinity College; Hon. Fellow of Emmanuel College. *Saxmeadham, Grange Road, Cambridge.*	1913–14
1910	Hopkinson, Bertram, M.A. M.Inst. C.E. Professor of Mechanism and Applied Mechanics in the University of Cambridge. 10 *Adams Road, Cambridge.*	
1900	Horne, John, LL.D. (Aberd. St. And.) F.R.S.E. F.G.S. Formerly Assistant Director of the Geological Survey. 12 *Keith Crescent, Blackhall, Midlothian.*	
1886	Horsley, Sir Victor Alexander Haden, Kt. B.S. F.R.C.S. M.D. (Halle); lately Professor of Pathology in University College, London. **Medal: Royal.** 25 *Cavendish Square,* W.; *and Athenæum Club.*	1898–99
1902	Hough, Sydney Samuel, M.A. (Camb.). His Majesty's Astronomer at the Cape of Good Hope. *Royal Observatory, Cape of Good Hope.*	
1893	Howorth, Sir Henry Hoyle, K.C.I.E. D.C.L. 45 *Lexham Gardens,* W.	
1889	Hughes, Thomas McKenny, M.A. Trin. Coll. Camb. F.G.S. F.S.A. Professorial Fellow of Clare College, Camb.; Chev. Ord. SS^{rum}. Maur. et Lazar. Ital.; Corresp. Memb. Soc. Géol. de Belg. and Soc. Géol. de Fr.; Woodwardian Professor of Geology in the University of Cambridge. *Ravensworth, Brooklands Avenue, Cambridge.*	
1867	Hull, Edward, M.A. LL.D. (Glasg.) F.G.S.; lately Director of the Geological Survey of Ireland, and Professor of Geology in the Royal College of Science; Master in Engineering (Hon. Caus. Dubl.); Hon. Mem. Acad. Sci. Amer. Philad., Soc. Géol. Belg., Geol. Soc. Edin., Glasg., Manch., Amer. Geog. Soc. 14 *Stanley Gardens, Notting Hill,* W.	
1906	Iveagh, Edward Cecil Guinness, Viscount, K.P. Hon. LL.D. (Dubl.). Chancellor of the University of Dublin. 5 *Grosvenor Place,* S.W.; *and Athenæum Club.*	
1901	Jackson, Sir Henry Bradwardine, K.C.B. K.C.V.O. Admiral; First Sea Lord of the Admiralty. *The Mall House, Admiralty,* S.W.	
1885	Japp, Francis Robert, M.A. LL.D. (St. And.) F.I.C. F.C.S. Emeritus Professor of Chemistry in the University of Aberdeen. 36 *Twyford Avenue, West Acton,* W.	
1906	Jeans, James Hopwood, M.A. Formerly Fellow of Trinity College, Cambridge; sometime University Lecturer in Mathematics, Cambridge; lately Professor of Applied Mathematics in Princeton University. 8 *Ormonde Gate, Chelsea,* S.W.	1914–15
1892	Joly, John, M.A. D.Sc. F.G.S. Professor of Geology and Mineralogy in the University of Dublin. **Medal: Royal.** *Somerset House, Temple Road, Rathmines, Dublin.*	1907–08

Date of Election		Service on Council, &c.
1877	Judd, John Wesley, C.B. LL.D. F.G.S. Emeritus Professor of Geology in the Royal College of Science, London, and Dean of the College; Soc. Phil. Ebor., Sci. Nat. Deva., Soc. Reg. Sydney, Soc. Asiat. Beng., Soc. Géol. du Nord, France, Soc. Belg. de Géol. de Pal. et d'Hydrol., Soc. Honor.; Acad. Sci. Nat. Philad., Soc. Géol. Belg. Brux., Inst. Imp. Geol. Vindob. Corresp.; Geol. Soc. Edin., Glasg. Liverp. Yorkshire, Soc. Emertius Professor of Geology in the Imperial College of Science and Technology. 30 *Cumberland Road, Kew, Surrey.*	1887–89 1902–04 **V.P.** 1902–04
1913	Keeble, Frederick William, M.A. Sc.D. (Camb.). Director Royal Horticultural Gardens, Wisley. *Weyton, St. George's Hill, Weybridge.*	
1913	Keith, Arthur, M.D. LL.D. (Aberd.) F.R.C.S. Hunterian Professor and Conservator of the Museum of the Royal College of Surgeons. 17 *Aubert Park, Highbury,* N.	
1881	Kempe, Sir Alfred Bray, Kt. M.A. D.C.L. (Dunelm)—**Treasurer** and **Vice-President**—2 *Paper Buildings, Temple,* E.C., *and* 22 *Pembridge Square,* W.	1897– **Treas.** 1898–
1887	Kennedy, Sir Alexander B. W., Kt., Emeritus Professor of Engineering in University College, London; LL.D. (Glasg. Birm.); D. Eng. (Liverp.); Hon. Fellow Roy. Soc. Edin., Hon. Mem. Roy. Soc. N.S.W., Mem. Inst. C.E., Hon. Mem. Inst. M.E. A7 *Albany, Piccadilly,* W.; *and Athenæum Club.*	1895–96
1909	Kerr, John Graham, M.A. F.R.S.E. F.L.S. F.Z.S. Regius Professor of Zoology, University of Glasgow; Formerly Fellow of Christ's College, Cambridge. 9 *The University, Glasgow.*	
1902	Kidston, Robert, LL.D. F.R.S.E. F.G.S. For. Mem. K. Mineral. Gesell. St. Petersburg. Corr. Mem. Soc. Géol. de Belgique, 12 *Clarendon Place, Stirling, N.B.*	
1897	Kipping, F. Stanley, D.Sc. (Lond.) Ph.D. (Munich). Professor of Chemistry, University College, Nottingham. *University College, Nottingham.*	1910–12
1887	Kirk, Sir John, G.C.M.G. K.C.B. M.D. LL.D. D.C.L. (Oxon) Sc.D. (Camb.) F.L.S. F.R.G.S. *Wavertree, Sevenoaks, Kent; and Athenæum Club.*	1893–95 **V.P.** 1894–95
1875	Klein, Edward Emanuel, M.D. Lately Lecturer on Advanced Bacteriology in the Medical School, St. Bartholomew's Hospital. *Lyndhurst, Willow Grove, Chislehurst, Kent.*	1888–90
1884	Lamb, Horace, M.A. Sc.D. (Camb. Dubl.) D.Sc. (Oxon) LL.D. (Glasg., St. Andr.) Professor of Mathematics in the University of Manchester. **Medal: Royal.** 6 *Wilbraham Road, Fallowfield, Manchester.*	1894–96 1908–10 **V.P.** 1909–10
1905	Lamplugh, George William, F.G.S. F.R.G.S. Assistant Director of the Geological Survey of Great Britain. 13 *Beaconsfield Road, St. Albans, Herts.*	1914–

Date of Election | Service on Council, &c.

1911 Lang, William Henry, M.B. C.M. D.Sc. (Glasg.) F.L.S. Barker Professor of Cryptogamic Botany in the University of Manchester. *University, Manchester.*

1883 Langley, John Newport, M.A. Sc.D. (Camb. Dubl.) Hon. LL.D. 1897–98
(St. Andr.); Hon. M.D. (Groningen). Professor of Physiology in 1904–05
the University of Cambridge; Fellow of Trinity College; Corr. **V.P.**
Mem. Soc. de Biol., Paris, Senck. Naturf. Gesell. Frankf., Royal 1904–05
Military Acad. of Med., Petrograd, K.B. Akad. d. Wissen,
Munich, and Gesell.-Deutsche Nervenærtze; Hon. Mem. K. K.
Gesell. d. Aerzte, Vienna, Soc. Aliéniste et Neurol. Univ. Imp..
Kazan., Soc. Med Suecana., American Physiol. Soc., For. Mem.
K. Vetensk. Soc., Upsala., K. Svenska Vetensk.-Akad., Stockholm,
K. Danske Videns. Selskab, Copenhagen, Acad. Roy. Méd. Belg.
Medal: Royal. *Hedgerley Lodge, Madingley Road, Cambridge;
and Athenæum Club.*

1875 Lankester, Sir Edwin Ray, K.C.B. M.A. and D.Sc. (Oxon) LL.D. (St. 1882–83
And.) D.Sc. (Leeds) Lately Director of the Natural History Depart- 88–90
ments, British Museum; lately Fullerian Professor of Physiology 94–96
in the Royal Institution; lately Linacre Professor of Comparative 1905–07
Anatomy and Fellow of Merton College, Oxford; Honorary **V.P.**
Fellow of Exeter College, and Honorary Student of Christ Church, 1895–96
Oxford; Emeritus Professor of Zoology in University College, 82–83
London; Mem. Inst. Fr. (Assoc. Étrang. Acad. Sci.); Hon. Fellow
Roy. Soc. Edin.; Hon. Mem. Camb. Phil. Soc., Roy. Phys. Soc.
Edin., Soc. de Biol. Paris, New York Acad. Sci. and Cali. Acad.
Sciences, San Francisco; Corr. Mem. Acad. Imp. Sci. Petrograd,
Roy. Acad. Sci. Turin, Acad. Nat. Sci. Philadelphia and
Senckenberg, Natur. Gesell. Frankfurt; For. Mem. R. Accad.
dei Lincei, Böhm. Gesell. Wiss., Königl. Gesell. Wiss. Gött., Amer.
Acad. Arts and Sci., Boston; For. Assoc. U. S. Nat. Acad. Sci.;
Mem. Amer. Phil. Soc.; Assoc. Acad. Roy. Belg. **Medals: Copley,
Royal.** 29 *Thurloe Place,* S.W.

1910 Lapworth, Arthur, D.Sc. (Lond.) B.Sc. (Birm.) F.I.C. Professor of Organic Chemistry in the University of Manchester. *University, Manchester; and 26 Broadway, Withington, Manchester.*

1888 Lapworth, Charles, LL.D. (Aberd.), F.G.S. Lately Professor of 1895–97
Geology in the University of Birmingham. **Medal: Royal.**
38 *Calthorpe Road, Edgbaston, Birmingham.*

1892 Larmor, Sir Joseph, Kt. M.A. (Camb.) D.Sc. (Lond. Dub. Oxon). 1897–99
LL.D. (Glasg. Aberd. Birm. St. And.) D.C.L. (Durh.) Hon. 1901–14
F.R.S. Edin., F.R.A.S. F.C.P.S., Hon. Mem. Amer. Acad. Arts and **Sec.**
Sci., Washington Acad. Sci., Lit. and Phil. Soc., Manch.; For. 1901–12
Mem. American National Acad. of Sci., R. Accad. dei Lincei, Rome; **V.P.**
Corr. Mem. Inst. di Bologna. Lucasian Professor of Mathematics 1912–14
in the University of Cambridge. Fellow of St. John's College,
Cambridge; formerly Professor of Natural Philosophy in Queen's
College, Galway, and Fellow of the Royal University of Ireland.
Medal: Royal. *St. John's College, Cambridge.*

Date of Election		Service on Council, &c.
1911	Leathes, John Beresford, **M.A. M.B.** (Oxon) **F.R.C.S.** Professor of Physiology in the University of Sheffield. *University, Sheffield.*	
1906	Lees, Charles Herbert, D.Sc. Professor of Physics in the University of London (East London College), Mile End, E. *Greenacres, Woodside Road, Woodford Green, Essex.*	1913–14
1910	Leishman, Sir William Boog, Kt. M.B. C.M. Colonel, R.A.M.C., Professor of Pathology, Royal Army Medical College. *Royal Army Medical College, Grosvenor Road, S.W., and Rotherwood, Linden Gardens, Leatherhead.*	
1909	Lewis, William James, M.A. (Oxon. Cantab.) F.R.A.S. F.C.S. Fellow of Oriel College, Oxford; Professor of Mineralogy in the University of Cambridge. *Trinity College, Cambridge.*	
1898	Lindley, Right Hon. Nathaniel, Lord. Serjeant-at-Law. LL.D. (Camb. Edin.), D.C.L. (Oxon.); late Master of the Rolls and Lord of Appeal. *The Lodge, East Carleton, Norwich; and Athenæum Club.*	
1900	Lister, Joseph Jackson, M.A. F.L.S. F.Z.S. *St. John's College, and Merton House, Grantchester; Cambridge.*	
1879	Liveing, George Downing, M.A. Sc.D. (Camb. Dubl.). Formerly Professor of Chemistry in the University of Cambridge; Fellow and President of St. John's College, Cambridge. **Medal: Davy.** *Newnham, Cambridge.*	1891–92 1902–04 **V.P.** 1891-9 1903–04
1882	Liversidge, Archibald, M.A. (Camb.) LL.D. (Glasg.) Assoc. R.S.M.; Hon. F.R.S.Edin. F.I.C. F.C.S. F.G.S. F.R.G.S.; Emeritus Professor of Chemistry in the University of Sydney; Memb. Phil. Soc. Camb., Phys. Soc. Lond., Min. Soc. Gr. Brit., Min. Soc. France; Hon. Mem. Roy. Soc. N.S.W., Vict., New Zeal. Inst., Roy. Histor. Soc., and K. Leop.-Carol. Akad. Halle; Corr. Mem. New York Acad. Sci., Senck. Naturf. Gesell. Frankf., Roy. Soc. Tasm., Roy. Soc. Queensland, Soc. d'Acclimat. Maur., Edin. Geol. Soc. *Fieldhead, Coombe Warren, Kingston, Surrey; and Athenæum Club.*	
1869	Lockyer, Sir J. Norman, K.C.B. LL.D. (Aberd. Edin. Glasg.) Sc.D. (Camb.) D.Sc. (Oxon. Sheffield) Hon. Mem. Phys. Soc. Lond., Ord. Imp. Bras. Rosae. Eq. Inst. Fr. (Acad. Sci.), Soc. pro fov. Indust. Nat. Par., Soc. Reg. Sci. Gött., Frank. Inst. Philad., Soc. Phys., Soc. Reg. Med. Brux., Soc. Spettros. Ital., Reg. Acad. Panorm., Bataaf. Genoots. Roterod. et Hist. Nat. Genev. Mem. Corr.; Acad. Reg. Linc. Romæ, Petrop, et Soc. Phil. Amer. Philad. Soc. Astron. France, Socius.; Soc. Lit. et Phil. Manc., Acad. Gioen. Sci. Nat. Catan., Soc. Phil. Ebor. Acad. Sci. New York, Soc. Astron. Mexico, et Univ. Lehigh Soc. Honor. **Medal: Rumford.** *Hill Observatory, Salcombe Regis, Devon.*	1874–76 1885–87 1891–93 **V.P.** 1892–93

Date of Election		Service on Council, &c.
1887	Lodge, Sir Oliver Joseph, Kt. D.Sc. (Oxon, Lond. Vict.) Sc.D. (Camb.) LL.D. (Glasg. St. And. Aberd.); Principal of the University of Birmingham; M. Inst. E.E.; Hon. Mem. Lit. Phil. Soc. Manch.: Corr. Mem. Amer. Phil. Soc. Philad., Accad. Sci. dell' Istituto Bologna, Bataafsch Genoots, Rotterdam. **Medal: Rumford.** *Mariemont, Edgbaston, Birmingham.*	1893–94
1902	Long, Right Hon. Walter Hume. 65 *Eaton Square*, S.W.	
1894	Love, Augustus Edward Hough, M.A. D.Sc. Sedleian Professor of Natural Philosophy in the University of Oxford, and Hon. Fellow of Queen's College. **Medal: Royal.** 34 *St. Margaret's Road, Oxford.*	1902–04 1914–
1914	Lowry, Thomas Martin, D.Sc. (Lond.). 17 *Eliot Park, Lewisham*, S.E.	
1913	Lucas, Keith, M.A. Sc.D. (Camb.). Lecturer in Physiology, and Fellow of Trinity College, Cambridge. *Trinity College, Cambridge.*	
1912	Lyle, Thomas Ranken, M.A. Sc.D. (Dubl.) Professor of Natural Philosophy in the University of Melbourne. *Lisbuoy, Toorak, Melbourne, Australia.*	
1906	Lyons, Henry George, Major R.E. (retired) Hon. D.Sc. (Oxon) Hon. Sc.D. (Dubl.) F.G.S. F.R.A.S. F.R. Met. Soc. Keeper Science Museum, South Kensington, S.W. First Class (Grand Cordon) Order Medjidieh, Second Class Order Osmanieh; Hon. Mem. Imp. Archæol. Inst. Berlin, Corresp. Mem. German Meteorol. Soc., Gesell. für Erdkunde, Berlin. 3 *Durham Place, Chelsea*, S.W.	
1881	Macalister, Alexander, M.A. M.D. (Dubl. and Camb.) Sc.D. (Dubl.) LL.D. (Glasg. Edin. and Univ. McGill) Professor of Anatomy in the University of Cambridge. *Torrisdale, Cambridge.*	1894–95
1906	Macallum, Archibald Byron, M.A. M.B. (Toronto) Ph.D. (Johns Hopkins) Sc.D. (Dubl. and Yale) LL.D. (Aberd. and McGill) F.R.S. (Canada); Hon. Mem. Roy. Phil. Soc. Glasg.; Assoc. Fellow (Hon.) of the Coll. Phys. Philad.; Professor of Biochemistry in the University of Toronto. *The University, Toronto, Canada.*	
1905	MacBride, Ernest William, M.A. (Camb.) D.Sc. (Lond.) LL.D. (McGill). Professor of Zoology in the Imperial College of Science and Technology, South Kensington. Formerly Strathcona Professor of Zoology in McGill University. 70 *Palace Gardens Terrace, Kensington*, W.	1912–14
1909	McClelland, John Alexander, M.A. (Camb.) M.A. and Hon. D.Sc. (National Univ. Ireland). Professor of Experimental Physics, University College, Dublin; Commissioner of National Education, Ireland; Secretary, Royal Irish Academy. *Rostrevor, Orwell Road, Rathgar, Dublin.*	
1901	Macdonald, Hector Munro, M.A. Professor of Mathematics in the University of Aberdeen. *University, Aberdeen.*	1908–10

Date of Election		Service on Council, &c.
1888	Macdonald, Right Hon. Sir John Hay Athole, K.C.B. LL.D. F.R.S.E. M.I.E.E. Lately Lord Justice-Clerk of Scotland, and Lord President of the Second Division of the Court of Session. 15 *Abercromby Place, Edinburgh.*	
1912	McDougall, William, M.A. M.B. (Camb.) M.Sc. (Manch.) Wilde Reader in Mental Philosophy in the University of Oxford. *Woodsend, Foxcombe Hill, Oxford.*	
1895	Macewen, Sir William, Kt. M.D. (Glasg.) Hon. LL.D. (Glasg. Liverp.) Hon. D.Sc. (Oxon Dubl.) Hon. F.R.C.S. Ireland. Regius Professor of Surgery in the University of Glasgow. 3 *Woodside Crescent, Glasgow.*	
1877	M^c^Intosh, William Carmichael, M.D. (Edin.) LL.D. (St. And. Edin.) F.L.S. F.R.S.E. L.R.C.S.E. C.M.Z.S. Professor of Natural History in the University of St. Andrews; Director of the University Museum, and of the Gatty Marine Laboratory, St. Andrews; Pres. Ray Soc.; Mem. Lit. and Antiq. Soc. Perth; Hon. Mem. Roy. Zool. Soc. Ireland, and Nat. Hist. Soc. Glasgow; Hon. Fell. Scot. Nat. Hist. Soc.; Hon. Mem. Psychol. Soc. Paris, and Soc. Centrale d'Aquicult. de France. **Medal: Royal.** 2 *Abbotsford Crescent, St. Andrews, N.B.; and Nevay Park, Meigle, Perthshire.*	1900–01
1884	M^c^Kendrick, John Gray, M.D. LL.D. F.R.S.E. F.R.C.P.E. Emeritus Professor of Physiology in the University of Glasgow. *Maxieburn, Stonehaven, N.B.*	1892–93
1915	Mackenzie, Sir James, M.D. F.R.C.P. LL.D. 133 *Harley Street*, W.	
1915	M^c^Lennan, John Cunningham, Ph.D. (Toronto) Professor of Physics and Director of the Physical Laboratory, University of Toronto. *University, Toronto; and* 88 *Prince Arthur Avenue, Toronto.*	
1881	M^c^Leod, Herbert, LL.D. (Hon. St. And.), F.C.S. Formerly Professor of Chemistry in the Royal Indian Engineering College, Cooper's Hill. 37 *Montague Road, Richmond, Surrey.*	1887–89
1890	MacMahon, Percy Alexander, Major, R.A. (retired) D.Sc. (Dubl.) Sc.D. (Camb.) LL.D. (Aberd. St. And.) F.R.A.S. Hon. Mem. C.P.S.; Deputy Warden of the Standards, Board of Trade. **Medal: Royal.** 27 *Evelyn Mansions, Carlisle Place*, S.W.; 7 *Old Palace Yard, Westminster*, S.W.; *and Athenæum Club.*	1895–97 1909–11 1915–
1908	Major, Charles Immanuel Forsyth, M.D. (Basle) F.Z.S. 10 *Via Venezia, Florence, Italy.*	
1903	Mallock, Henry Reginald Arnulph. 3 *Victoria Street*, S.W.; *and* 6 *Cresswell Gardens, South Kensington*, S.W.	1910–12
1900	Manson, Sir Patrick, G.C.M.G. M.D. (Aberd.) LL.D. (Aberd.) F.R.C.P. Hon. D.Sc. (Oxon). Lately Physician and Medical Adviser to the Colonial Office; lately Lecturer on Tropical Medicine to St. George's Hospital, Charing Cross Hospital, and London School of Tropical Medicine. *The Sheiling, Clonbur, Co. Galway.*	

Date of Election		Service on Council, &c
1873	*Markham, Sir Clements Robert, K.C.B. F.R.G.S. F.S.A. Acad. Caes. Nat. Cur. Socius; Soc. Geog. Par., Berol., Vindob., Hist. Philad. et Univ. Chil. Soc. Honor. *Athenæum Club; and 21 Eccleston Square*, S.W.	
1891	Marr, John Edward, M.A. Sc.D. (Camb.) F.G.S. Fellow and Lecturer of St. John's College, Cambridge, and University Lecturer in Geology. *St. John's College, Cambridge.*	1904–06
1906	Marsh, James Ernest, M.A. (Oxon) F.C.S. Fellow of Merton College. *University Museum; and 2 Davenant Road, Oxford.*	
1901	Martin, Charles James, M.D. D.Sc. (Lond.); Hon. D.Sc. (Dubl. Melbourne, Sheffield). Director of the Lister Institute; Professor of Experimental Pathology in the University of London. *Lister Institute of Preventive Medicine, Chelsea Bridge Road*, S.W.	1909–11
1895	Martin, Sidney Harris Cox, M.D. B.S. B.Sc. F.R.C.P. Physician to University College Hospital. 51 *Wimpole Street*, W.	
1903	Masson, David Orme, M.A. D.Sc. (Edin.). Professor of Chemistry, University of Melbourne. *University of Melbourne, Victoria, Australia.*	
1915.	Masterman, Arthur Thomas, M.A. (Camb.) D.Sc. (Lond. St. And.) F.R.S.E. Superintending Inspector, H.M. Board of Agriculture and Fisheries. 43 *Parliament Street*, S.W.	
1902	Mather, Thomas, M.Inst.E.E. Professor of Electrical Engineering in the City and Guilds (Engineering) College, Imperial College of Science and Technology, South Kensington. 37 *Wyatt Park Road, Streatham Hill*, S.W.	
1897	Mathews, George Ballard, M.A. Formerly Fellow of St. John's College, Cambridge; Fellow of University College, London; lately Professor of Mathematics, and now Special Lecturer, in the University College of North Wales. 10 *Menai View, Bangor, N.W.*	1905–06
1898	Maxwell, Right Hon. Sir Herbert Eustace, Bart., D.C.L. LL.D. F.S.A. Pres. Soc. Antiq. Scott. 1 *St. James's Place*, S.W.; *Carlton Club; and Monreith, Whauphill, Wigtownshire, N.B.*	
1912	Messel, Rudolph, Ph. D. (Tüb.) F.C.S. 147 *Victoria Street, Westminster*, S.W.	
1904	Meyrick, Edward, B.A. (Camb.) F.Z.S. *Thornhanger, Marlborough.*	
1892	Miall, Louis Compton, Lately Professor of Biology in the University of Leeds. 21 *Norton Way*, N., *Letchworth, Herts.*	
1902	Michell, John Henry, M.A. Lecturer on Mixed Mathematics in the University of Melbourne. *The University, Melbourne.*	

* Deceased, January, 1916.

Date of Election		Service on Council, &c.
1896	Miers, Sir Henry Alexander, Kt., D.Sc. (Oxon, Sheffield) M.A. (Oxon) Hon. Ph.D. (Christiania) F.G.S. F.C.S. Vice-Chancellor of Victoria University, Manchester. Mem. Min. Soc.; Corr. Mem. K.B. Akad. Wissen. Munich; Hon. Mem. Accad. d. Zelanti, Acireale, Soc. Linn. Normandie, Roy. Cornwall Polytech. Soc.; Fellow, Magdalen College, Oxford, and Eton College; lately Waynflete Professor of Mineralogy in the University of Oxford; Lately Principal of the University of London. *Birch Heys, Cromwell Range, Fallowfield, Manchester.*	1901-03
1874	Mills, Edmund James, Hon. LL.D. (Glasg.) D.Sc. (Lond.) F.C.S. F.I.C. Corr. Mem. Roy. Phil. Soc. Glasg. Emeritus Professor of Technical Chemistry in the Royal Technical College, Glasgow. 64 *Twyford Avenue, West Acton*, W.	
1906	Mitchell, Peter Chalmers, M.A. (Oxon, Aberd.) D.Sc. (Oxon) LL.D. (Western Univ. Pennsyl.) F.L.S. F.Z.S. Secretary of the Zoological Society of London. Hon. Mem. Roy. Zool. Soc. Ireland; Corr. Mem. Acad. Sci. Philad. *Zoological Society, Regent's Park*, N.W.	
1912	Moore, Benjamin, B.E. M.A. D.Sc. (Queen's Univ. Belfast) M.R.C.S. L.R.C.P. Research Staff, Department of Applied Physiology, Medical Research Committee, National Insurance Commission. Lately Professor of Biochemistry, University of Liverpool. Formerly Professor of Physiology, Yale University, New Haven, U.S.A. 8 *Pembroke Villas, The Green, Richmond*, S.W.	
1899	Morgan, Conwy Lloyd, LL.D. (Aberd.) D.Sc. (Wales and Bristol), A.R.S.M. F.G.S. Professor of Psychology in the University of Bristol. Corr. Acad. Sci. New York and Philad. 5 *Kensington Place, Clifton, Bristol.*	
1915.	Morgan, Gilbert Thomas, D.Sc. (Lond.) F.I.C. F.C.S. M.R.I.A. Assoc. Roy. Coll. Sci. (Lond.). Professor of Chemistry, Royal College of Science for Ireland; formerly Assistant Professor in the Royal College of Science, London (Imperial College of Science and Technology). *Royal College of Science, Dublin.*	
1892	Morley of Blackburn, Right Hon. John, Viscount, O.M. M.A. D.C.L. (Oxon) LL.D. (Camb. and Glasg.) Trust. Brit. Mus. *Flowermead, Wimbledon Park; and Athenæum Club.*	
1896	Mott, Frederick Walker, M.D. (Lond.) F.R.C.P. Formerly Fullerian Professor of Physiology in the Royal Institution. *Pathological Laboratory, Claybury Asylum, Essex; and* 25 *Nottingham Place*, W.	1908-09
1880	Moulton, Right Hon. Lord, of Bank, K.C.B. LL.D. Lord of Appeal. 57 *Onslow Square*, S.W.	
1911	Muir, Robert, M.A. M.D. Sc.D. F.R.C.P. (Edin.). Professor of Pathology in the University of Glasgow. 16 *Victoria Crescent, Glasgow.*	

Date of Election		Service on Council, &c.
1900	Muir, Sir Thomas, Kt. C.M.G. M.A. LL.D. F.R.S.E. Lately Superintendent-General of Education in Cape Colony. *Elmcote, Sandown Road, Rondebosch, South Africa.*	
1904	Muirhead, Alexander, D.Sc. (Lond.). *The Lodge, Shortlands, Kent.*	
1915	Myers, Charles Samuel, M.D. Sc.D. (Camb.). Director of the Psychological Laboratory, and Lecturer in Experimental Psychology in the University of Cambridge. *Great Shelford, Cambridge.*	
1908	Nevill, Edmund Neville, F.R.A.S. Director of the Government Observatory, Natal. *Ellaker, Enys Road, Eastbourne.*	
1902	Newall, Hugh Frank, M.A. (Camb.) D.Sc. (Durh.) F.R.A.S. Professor of Astrophysics and Director of the Solar Physics Observatory in the University of Cambridge. *Madingley Rise, Cambridge ; and Athenæum Club.*	1905–07
1912	Newstead, Robert, M.Sc. (Liverp.) A.L.S. F.E.S. Professor of Entomology in the School of Tropical Medicine, University of Liverpool. "*Witcombe,*" 67 *Handbridge, Chester.*	
1893	Newton, Edwin Tulley, F.G.S. F.Z.S. Hon. Mem. Norfolk Nat. Soc. *Florence House, Willow Bridge Road, Canonbury,* N.	
1880	Niven, Charles, M.A. D.Sc. Professor of Natural Philosophy in the University, Aberdeen. 6 *Chanonry, Old Aberdeen.*	
1882	Niven, Sir William Davidson, K.C.B. M.A. LL.D., lately Director of Studies in Royal Naval College, Greenwich. *Eastburn, St. John's Road Sidcup, Kent.*	1892–94 1904–06 V.P. 1904–06
1890	Norman, Rev. Alfred Merle, M.A. D.C.L. Hon. LL.D. (St. And.) F.L.S. Hon. Canon of Durham. *The Red House, Berkhamsted, Herts.*	
1900	Northumberland, Henry George Percy, Duke of, K.G. F.S.A. President of the Royal Institution. 2 *Grosvenor Place,* S.W.; *and Athenæum Club.*	1910–12 V.P. 1910–12
1904	Nuttall, George Henry Falkiner, M.A. (Camb.) Sc.D. (Camb.) M.D. (Cal.) Ph.D. (Gött.). Corr. Mem. Roy. Acad. Med. Turin. Quick Professor of Biology in the University of Cambridge. "*Longfield,*" *Madingley Road, Cambridge.*	
1859	Odling, William, M.A. (Oxon) M.B. (Lond.) Coll. Reg. Med. Socius. V.P.C.S. Hon. Math. Phys. Doct. (Lugd. Bat.) Lately Waynflete Professor of Chemistry in the University of Oxford. 15 *Norham Gardens, Oxford.*	1864–66 79–81

Date of Election		Service on Council, &c.
1911	Oldham, Richard Dixon, F.G.S. M.Inst. M.M., A.R.S.M. 8 *North Street, Horsham, Sussex.*	
1863	Oliver, Daniel, LL.D. (Aberd.) F.L.S. Lately Keeper of the Herbarium and Library, Royal Gardens, Kew; Emeritus Professor of Botany, University College, London. **Medal: Royal.** 10 *Kew Gardens Road, Kew.*	1875–76 80–82
1905	Oliver, Francis Wall, M.A. (Camb.) D.Sc. (Lond.) F.L.S. Professor of Botany in the University of London. *University College, Gower Street,* W.C.	
1912	Oram, Sir Henry John, K.C.B. Engineer Vice-Admiral; Engineer-in-Chief of the Fleet. M.Inst.C.E. M.Inst.M.E. *Admiralty, Whitehall,* S.W.; *and Athenæum Club.*	
1909	Orr, William McFadden, M.A. (Camb. Queen's Univ. Belfast). Professor of Mathematics, Royal College of Science for Ireland. *Innisfail, Bushy Park Road, Terenure, Dublin.*	
1898	Osler, Sir William, Bart. M.D. F.R.C.P. Hon. D.Sc. (Oxon). Regius Professor of Medicine in the University of Oxford. 13 *Norham Gardens, Oxford.*	
1882	Palgrave, Sir Robert Harry Inglis, Kt. F.S.S. Knight, Order of Wasa of Sweden; M.Inst. Internat. de Statistique, Stat. and Social Inquiry Society of Ireland, Soc. d'Économie Politique; Hon. Mem. Lit. Phil. Soc., Manchester. *Henstead Hall, Wrentham, Suffolk.*	
1898	Parsons, Hon. Sir Charles Algernon, K.C.B. M.A. D.Sc. (Camb. Oxford, Dublin, Durham) LL.D. (Glasgow) D.Eng. Liverpool; Hon. Mem. Inst. Ingén. Civil de France, North E. Coast Inst. of Engineers and Shipbuilders, Inst. Mech. Engineers; Hon. Fellow, St. John's Coll. Camb. **Medal: Rumford.** 6 *Windsor Terrace, Newcastle-on-Tyne.*	1907–09 **V.P.** 1908–09
1914	Paton, Diarmid Nöel, M.D. F.R.C.P. (Edin.) F.R.S.E. Professor of Physiology in the University of Glasgow. *University, Glasgow.*	
1892	Peach, Benjamin Neeve, F.R.S.E. F.G.S. *Geological Survey Office,* 33 *George Square, Edinburgh.*	
1896	Pearson, Karl, M.A. LL.B. Hon. LL.D. (St. And.) Galton Professor of Eugenics, University of London; Hon. Fellow of King's College, Cambridge; Hon. Mem. Anthrop. Soc. St. Petersburg, Washington. **Medal Darwin.** 7 *Well Road, Hampstead,* N.W.	
1892	Pedler, Sir Alexander, Kt. C.I.E. F.C.S. F.I.C. Lately Vice-Chancellor of the University of Calcutta, and Director of Public Instruction with the Government of Bengal. 28 *Stanhope Gardens, Queen's Gate,* S.W.	
1903	Perkin, Arthur George, F.C.S. F.I.C. F.R.S.E. *Grosvenor Lodge, Grosvenor Road, Headingley, Leeds.*	
1890	Perkin, William Henry, Ph.D. (Würzburg) Sc.D. (Camb.) LL.D. (Edin. St. Andr.) M.Sc. (Vict.) Corr. Mem. K. Gesell. Wissen. Gött., Königl. Bayer. Akad. Wissen. Munich. F.I.C. F.C.S. Waynflete Professor of Chemistry in the University of Oxford. **Medal: Davy.** 5 *Charlbury Road, Oxford.*	1904–05 1908–10

Date of Election		Service on Council, &c.
1885	Perry, John, D.Sc. LL.D. M.E. (Q. Univ. Ireland) D.Sc. (R. Univ. Ireland) LL.D. (Glasg.) D.Sc. (Syd.). Doctor of Eng. Educ., Japan; Emeritus Professor of Mechanics and Mathematics in the Imperial College of Science and Technology, London. 25 *Stanley Crescent, Notting Hill*, W.	1904–06
1907	Petavel, Joseph Ernest, D.Sc. (Manch.) Professor of Engineering in the University of Manchester, and Director of the Whitworth Laboratory. *The University, Manchester.*	
1902	Petrie, William Matthew Flinders, Hon. D.C.L. (Oxon) Litt. D. (Camb.) LL.D. (Edin. Aberd.) Hon. F.S.A. (Scot.) Mem. Roy. Anthrop. Inst., Roy. Irish Acad., Amer. Phil. Soc., Imp. German Arch. Inst., Ital. Soc. Anthrop., Roman Soc. Anthrop., Soc. Northern Antiq.; Corr. Mem. Soc. Anthrop. Berlin, Soc. Arch. France, Numis. Soc. Philad. Professor of Egyptology in London University. 8 *Well Road, Hampstead*, N.W.	
1887	Pickard-Cambridge, Rev. Octavius, M.A. *Bloxworth, Wareham, Dorset.*	
1890	Pickering, Spencer Percival Umfreville, M.A. F.C.S. F.I.C. Mem. Phys. Soc. Lond. *Harpenden, Herts; Woolacombe, N. Devon.*	
1910	Plimmer, Henry George, M.R.C.S. F.L.S. F.Z.S. F.R.M.S. Mem. Deutsch. Pathol. Gesell. Professor of Comparative Pathology in the Imperial College of Science and Technology, South Kensington, S.W. Pathologist to the Zoological Society; lately Pathologist and Lecturer on Pathology and Bacteriology, St. Mary's Hospital, London. 32 *Queen's Road, St. John's Wood*, N.W.	
1902	Plunkett, Right Hon. Sir Horace Curzon, K.C.V.O. M.A. Hon. D.C.L. (Oxon). Formerly Vice-President of the Department of Agriculture and Technical Instruction for Ireland. 105 *Mount Street*, W.	
1907	Pocklington, Henry Cabourn, M.A. (Camb.) D.Sc. (Lond.). 5 *Wellclose Place, Leeds.*	
1911	Pocock, Reginald Innes, F.L.S. F.Z.S. Superintendent of the Zoological Society's Gardens; formerly Senior Assistant in the Zoological Department of the British Museum. *Zoological Society, Regent's Park*, N.W.	
1902	Pope, William Jackson, M.A. (Camb.) M.Sc. (Manch.) Hon. LL.D. (St. And.) F.C.G.I. F.I.C. Fellow of Sidney Sussex College and Professor of Chemistry in the University of Cambridge. **Medal: Davy.** *University Chemical Laboratories, Cambridge.*	1910–11
1911	Porter, Alfred William, B.Sc. (Lond.). Fellow of, and Assistant Professor in, the University of London, University College. *University College, Gower Street*, W.C.	

Date of Election		Service on Council, &c.
1889	Poulton, Edward Bagnall, M.A. (Oxon.) D.Sc. (Oxon. Sydney), Hon. LL.D. (Princeton), Hon. D.Sc. (Durh.), Fellow of Jesus College, and Hope Professor of Zoology in the University of Oxford. Pres. L.S. F.Z.S. F.G.S. Commander (Cl. II) Swedish Order of the Pole Star; Hon. Memb. Soc. Ent. Belg., Real Soc. Españ. Hist. Nat., New York Acad. Sci.; Corresp. Mem. Acad. Nat. Sci. Philad., Soc. Nat. Hist. Boston, and American Ent. Soc. **Medal: Darwin.** *Wykeham House, Banbury Road, Oxford; St. Helen's Cottage, St. Helen's, Isle of Wight; and Athenæum Club.*	1897–99 1908–10 **V.P.** 1909–10
1895	Power, Sir William Henry, K.C.B., Lately Medical Officer to H.M. Local Government Board. **Medal: Buchanan.** *Holly Lodge, East Molesey, Surrey.*	
1905	Prain, Sir David, Lieut.-Colonel I.M.S. (retired) C.M.G. C.I.E. M.A. M.B. LL.D. (Aberd. St. Andr.) F.R.S.E. F.L.S. Director Royal Botanic Gardens, Kew; lately Director Botanical Survey of India, and Superintendent, Royal Botanic Gardens, Calcutta; For. Mem. K. Svenska Vetensk. Akad. Stockholm; Hon. Mem. Accad. d. Zelanti, Acireale, Deutsch. Bot. Gesell., Pharm. Soc. Gt. Britain; Corr. Mem. K.B. Akad. Wissen. Munich, Inst. Nat. Genev., Nederl. Bot. Vereen. Amsterdam, Soc. Nat. Sci. Cherbourg. *Royal Botanic Gardens, Kew, Surrey.*	1908–10
1912	Prior, George Thurland, M.A. D.Sc. (Oxon) F.G.S. Keeper of the Department of Mineralogy, British Museum (Natural History). *British Museum (Natural History), S. Kensington,* S.W.	
1912	Punnett, Reginald Crundall, M.A. Fellow of Gonville and Caius College, Professor of Biology in the University of Cambridge. *Caius College, Cambridge; Whittingehame Lodge, Cambridge.*	
1895	Purdie, Thomas, B.Sc. Ph.D. Hon. LL.D. (Aberd. St. Andr.) A.R.S.M. Emeritus Professor of Chemistry in the University of St. Andrews. *The University, St. Andrews.*	
1900	Rambaut, Arthur Alcock, M.A. (Dubl. Oxon) Sc.D. (Dubl.) F.R.A.S. M.R.I.A. F.R. Met. Soc. Radcliffe Observer. *Radcliffe Observatory, Oxford.*	
1888	Ramsay, Sir William, K.C.B., Commander of the Crown of Italy; Officier Lég. Hon. France; Knt. Pruss. Ord. "*Pour le Mérite*"; Ph.D. (Tüb., Cracow) LL.D. (Glasg. Johns Hopkins, Baltimore) Sc.D. (Dubl. Camb. Oxon. Liverp. Birm. Sheffield, Columbia, Bristol, Christiania, Durh.) M.D. (Heid., Jena), F.I.C. F.C.S.; Mem. Soc. Chem. Ind. Lately Professor of Chemistry, University of London; Nobel Laureate, Chemistry, 1904. For. Mem. Inst. France (Acad. Sci., and Acad. Méd.), Acad. Imp. Bohemia, K. Svenska Vetensk. Akad., Phil. Soc. Philad., Amer. Acad., Acad. Sci. Bologna, R. Accad. dei Lincei, Vidensk. Selskab, Christiania, Soc. Ital. (detta dei XL), R. Acad. of Netherlands; Hon. Mem. Soc. Phys. et Sci. Nat. Genève, Roy. Irish	1896–97

Date of Election		Service on Council &c.
	Acad., Pharm. Soc., Physikal. Verein, Frankfort, Deutsch. Chem. Gesell. Berlin, Chem. Soc. Belg., Soc. Sci. Roumania, Lit. Phil. Soc. Manch., New York Acad. Sci., Amer. Chem. Soc., Kongl. Danske Vidensk. Selskab, Soc. Hélvét. des Sci., Soc. Med. Erlangen, Bunsen Gesell., Brooklyn Inst., Roy. Soc. Edin., Roy. Med. Chir. Soc., Soc. Anthrop. Moscow, Soc. Española Phys.-Quim., Soc. Chim. de France, Roy. Cornwall Polytech. Soc., Phil. Soc. Glasg., Soc. Pub. Anal., Asiatic Soc. Bengal; Corresp. Mem. K. Akad. d. Wissen. Berlin, R. Accad. Turin, Imp. Acad. Sci. Petrograd, Acad. Roy. Hungary, R. Acad. Ciencias Madrid, Bat Genoots. Rotterdam, R. Istit. Veneto, Soc. São Paulo, Brazil, Inst. Ant. Alvarez Mexico, Oesterr. Ingen. u. Arch. Verein. **Medal: Davy.** *Beechcroft, Hazlemere, High Wycombe.*	
1884	Ransome, Arthur, M.A. M.D. F.R.C.P. Lately Professor of Public Health in Owens College, and Examiner in Sanitary Science in Cambridge and Manchester Universities. Hon. Fell. of Caius Coll., Cambridge. *Sunnyhurst, Dean Park, Bournemouth.*	
1873	Rayleigh, Right Hon. John William Strutt, Lord, O.M.—**Past President**—M.A. D.C.L. (Oxon) Sc.D. (Camb. and Dubl.) LL.D. (Edin. Glasg. Toronto, and Univ. McGill) Ph.D. (Heidel.) Chancellor of the University of Cambridge, and formerly Cavendish Professor of Physics; Hon. Fellow of Trinity College, Cambridge; Officier Lég. Hon. France; Nobel Laureate, Physics, 1904; Hon. Mem. Inst. C.E.; F.R.A.S.; Soc. Reg. Edin., Acad. Reg. Hib., Soc. Lit. et Phil. Manc., Acad. Reg., Sci. Monach., Soc. Asiat. Beng., Soc. Honor.; For. Mem. Inst. France (Acad. Sci.); Acad. Reg. Sci. Hafn., Soc. Reg. Sci. Gött., Acad. Sci. Berol., Acad. Imp. Sci. Petropol. Corr. Soc.; Scientific Adviser to the Trinity House; Hon. Professor of Natural Philosophy in the Royal Institution. **Medals: Copley, Royal, Rumford.** *Terling Place, Witham, Essex.*	1877–79 84–96 1905–08 **Sec.** 1885–96 **Pres.** 1905–08
1899	Reid, Clement, F.L.S. F.G.S. *One Acre, Milford-on-Sea, Hants.*	
1898	Reid, Edward Waymouth, B.A. M.B. Sc.D. (Camb.) Professor of Physiology in University College, Dundee, St. Andrews University. *University College, Dundee.*	
1883	Reinold, Arnold William, C.B. M.A. (Oxon). Lately Professor of Physics in the Royal Naval College, Greenwich, and Lecturer at Guy's Hospital; F. Phys. Soc. Lond.; Hon. Mem. Yorks. Phil. Soc. 11 *Spencer Road, Southsea.*	1899–01
1909	Rendle, Alfred Barton, M.A. (Camb.) D.Sc. (Lond.) F.L.S. Keeper of the Department of Botany, British Museum (Nat. Hist.) 28 *Holmbush Road, Putney,* S.W.	
1880	Reynolds, J. Emerson, M.D. Sc.D. (Dubl.) F.C.S. F.I.C. Mem. Soc. Chem. Ind. Lately Professor of Chemistry, University of Dublin. 3 *Inverness Gardens, Kensington,* W.	1900–02 **V.P.** 1901–02

Date of Election		Service on Council, &c.
1913	Richardson, Owen Willans, M.A. D.Sc., Wheatstone Professor of Physics, University of London, King's College. 4, *Cannon Place, Hampstead*, N.W.	
1911	Richmond, Herbert William, M.A. Fellow and Lecturer of King's College, Cambridge, and University Lecturer in Mathematics. *King's College, Cambridge.*	
1907	Ridley, Henry Nicholas, C.M.G. M.A. (Oxon) F.L.S., lately Director of Gardens and Forests, Straits Settlements. 7 *Cumberland Road, Kew Gardens.*	
1908	Rivers, William Halse Rivers, M.A. (Camb.) M.D. (Lond.) F.R.C.P. Lecturer in the Physiology of the Senses, University of Cambridge; Fellow of St. John's College, Cambridge. **Medal: Royal.** *St. John's College, Cambridge.*	
1899	Romer, Right Hon. Sir Robert, G.C.B. M.A. 21 *Queen's Gate Gardens*, S.W.; *and Athenæum Club.*	
1886	Rosebery, Right Hon. Archibald Philip Primrose, Earl of, K.G. K.T. D.C.L. Trust. Brit. Mus. Chancellor of the University of London. 38 *Berkeley Square*, W.; *and Dalmeny Park, Linlithgowshire.*	
1913	Rosenhain, Walter, B.A. (Camb.) B.C.E. D.Sc. (Melb.) Superintendent of the Department of Metallurgy and Metallurgical Chemistry in the National Physical Laboratory. *Warrawee, Combe Lane, Kingston Hill, Surrey.*	
1901	Ross, Sir Ronald, K.C.B. Major, I.M.S. (retired), Off. de l'Ordre de Leopold II. Hon. F.R.C.S. (Eng.), D.P.H., R.C.P.S. London. Nobel Laureate, Physiology and Medicine, 1902; Hon. D.Sc. Trin. Coll. Dublin, and Univ. Leeds; LL.D. Aberdeen; Corresp. Étranger, Acc. de Méd. Paris; Hon. Mem. Medical Societies of Stockholm, Ghent, Batavia, Manila; of Tropical Medical Societies of Paris and America; and of Royal Physiographical Society of Lund. Hon. M.D. (Carol. Inst.. Stockholm), Hon. M.D. (Athens); Corr. Estero, Accad. di Med., Turin; Assoc. Fellow, Coll. Phys., Philadelphia. **Medal: Royal.** 18 *Cavendish Square*, W.	1911–13 **V.P.** 1911–13
1911	Rothschild, Lionel Walter, Lord, Ph.D. (Giessen) Trust. Brit. Mus. 148 *Piccadilly*, W.; *Tring Park, Tring, Herts.*	
1914	Ruhemann, Siegfried, M.A. Ph.D. (Berl.) University Lecturer in Organic Chemistry, Cambridge. *The Elms, Adams Road, Cambridge.*	
1908	Russell, Hon. Bertrand Arthur William. Lecturer and lately Fellow of Trinity College, Cambridge. *Trinity College, Cambridge.*	

Date of Election		Service on Council, &c.
1903	Rutherford, Sir Ernest, Kt. M.A. D.Sc. (N.Z.) B.A. (Camb.) LL.D. (Wisconsin, Pennsyl., McGill, Birm., Edin.) D.Sc. (Dubl.) Ph.D. (Giessen); D.Phys. (Clark); Nobel Laureate, Chemistry, 1908; Hon. Mem. N.Z. Phil. Inst., Amer. Phil. Soc. Acad. Sci. St. Louis, K. Vetensk. Soc. Upsala; Corr. Mem. Roy. Soc. Gött., Acad. Sci. Munich, Imp. Acad. Sci. Vienna, Soc. Med. Erlangen; Foreign Assoc. Nat. Acad. Sci. Washington. Langworthy Professor of Physics, University of Manchester. **Medal: Rumford.** 17 *Wilmslow Road, Withington, Manchester.*	1910–12
1903	Sampson, Ralph Allen, M.A. (Camb.) D.Sc. Astronomer Royal for Scotland. *Royal Observatory, Edinburgh.*	1909–10
1878	Schäfer, Sir Edward Albert, Kt. M.R.C.S. Hon. LL.D. (Aberd., St. Andr., McGill) Hon. Sc.D. (Trin. Coll. Dubl., Camb.) Hon. M.D. (Berne, Groningen, Melb.); Professor of Physiology in the University of Edinburgh. Formerly Jodrell Professor of Physiology in University College, London, and Fullerian Professor of Physiology, Royal Institution. Corr. Mem. Soc. Biol. Paris, R. Accad. Sci. Istit. Bologna; Hon. Mem. Amer. Med. Assoc., Amer. Physiol. Soc., K. Vetensk. Soc. Upsala, K. Leop.-Carol. Deutsch. Akad. Halle, Roy. Phil. Soc. Glasgow. **Medal: Royal.** *Marly Knowe, North Berwick.*	1890–92 1902–03
1901	Schlich, Sir William, K.C.I.E. Ph.D. M.A. (Oxon.) F.L.S. Professor of Forestry, University of Oxford. Lately Inspector-General of Forests to the Government of India. 29 *Banbury Road, Oxford.*	
1879	Schuster, Arthur—**Secretary**—Ph.D. Sc.D. (Camb.) D.Sc. (Manch. and Calcutta) Dr. ès. Sc. (Geneva) LL.D. (St. Andr.) F.R.A.S. Mem. Inst. Elect. Eng.; Hon. Mem. Roy. Phil. Soc. Glasg., Phil. Soc. Camb., Schweiz. Natur. Gesell., Med. Phys. Soc. Erlangen; For. Mem. Soc. d. Spettros. Ital., For. Assoc. Nat. Acad. Sci. Washington; Corr. Mem. Roy. Soc. Sci. Gött. Honorary Professor of Physics in the University of Manchester; President of the British Association. **Medal: Royal.** *Yeldall, Twyford, Berks.*	1885–87 98–99 1912– **Sec.** 1912–
1898	Scott, Alexander, Sc.D. (Camb.) D.Sc. (Edin.) M.A. (Durh.) F.R.S.E., Pres. Chem. Soc., F.I.C. Formerly Superintendent of the Davy Faraday Research Laboratory. 34 *Upper Hamilton Terrace,* N.W.	1906–08
1894	Scott, Dukinfield Henry—**Foreign Secretary**—M.A. (Oxon) D.Sc. (Manch.) LL.D. (Aberd.) Ph.D. (Würzb.) F.L.S. F.G.S. F.R.M.S. Lately Honorary Keeper of the Jodrell Laboratory, Royal Botanic Gardens, Kew. Hon. Mem. Lit. Phil. Soc. Manch., Roy. Botan. Soc., Botan. Soc. Edin., Mem. K. Danske Videns. Selskab, Copenhagen, K. Vetens.-Soc., Upsala, Soc. Linn. Normandie; Corr. Mem. K. B. Akad. Wissen., Munich, Deutsch. Botan. Gesell. R. Inst. Veneto di Scienze, Lett. ed Arti., Hon. Mem. Soc. Helvét. Sci. Nat. **Medal: Royal.** *East Oakley House, Oakley, Hants; and Athenæum Club.*	1897–99 1912– **For. Sec.** 1912–

Date of Election		Service on Council, &c
1870	Scott, Robert Henry, M.A. D.Sc. (Dubl.) F.Z.S. F.R. Met. Soc., lately Secretary to the Meteorological Council. Officer of the Legion of Honour; Ord. Coron. Ferr. Austr. Eq.; Acad. Cæs. Leop. Soc.; Soc. Met. Fr. Par., Soc. Imp. Reg. Zool. Bot., Soc. Met. Austr. Vindob., Soc. Met. Germ. Berol. et Soc. Nat. Scrutat. Emb. Soc. Honor.; Inst. Geol. Imp. Vindob. Soc. Met. Ital. Taurin. et Soc. Isis Dresd. Mem. Corr. 6 *Elm Park Gardens*, S.W.	1900-02
1905	Searle, George Frederick Charles, M.A. Sc.D. (Camb.), Fellow of Peterhouse, Cambridge. University Lecturer in Experimental Physics, and Demonstrator of Experimental Physics at the Cavendish Laboratory, Cambridge. *Wyncote, Hills Road, Cambridge.*	
1898	Seward, Albert Charles, M.A. (Camb.) F.G.S. F.L.S. Master of Downing College, Cambridge. Professor of Botany in the University of Cambridge, Fellow of St. John's College, Hon. Fellow of Emmanuel College. *Botany School; and Downing College, Cambridge.*	1906-08
1890	Sharp, David, M.B. C.M. (Edin.) Hon. M.A. (Camb.) F.Z.S. F.E.S. Hon. Mem. New Zealand Inst. and Entomological Societies of France, Germany and Russia. *Lawnside, Brockenhurst, Hants.*	
1891	Shaw, Sir William Napier, Kt. Sc.D. (Camb.) Hon. LL.D. (Aberd.) Hon. Sc.D. (Dubl., Manch.) Hon. D.Sc. (Harvard); Director of the Meteorological Office; Hon. Fellow Emmanuel College, Cambridge; Hon. Mem. Roy. Cornwall Polytechnic Soc., Meteorol. Soc. Mauritius, K.K. Oesterr. Gesell. für Meteorol., Deutsch. Meteorol. Gesell.; President of the International Meteorological Committee; Reader in Meteorology in the University of London. *Meteorological Office, South Kensington*, S.W.; 10 *Moreton Gardens*, S.W.; *and Athenæum Club.*	1903-05
1893	Sherrington, Charles Scott, M.A. M.D. (Camb.) M.A. (Oxon.) F.R.C.P. (Lond.) Sc.D. (Camb.) Hon. D.Sc. (Dubl. Sheffield) Hon. LL.D. (Glasgow, Toronto and Harvard); Waynflete Professor of Physiology in the University of Oxford; Honorary Fellow of Gonville and Caius College, Cambridge; Fellow of Magdalene College, Oxford; Mem. Corr. Honor. Soc. Neurol. Paris, Soc. Ital. di Neurol.; Ehrenmitglied d. Gesell. d. deutsch. Nervenaerzte; Mem. Corr. K. Med. Gesell. Vienna; Corr. Mem. Gesell. f. Inn. Med. Vienna, Physikal.-Mediz. Soc. Erlangen, Acad. Imp. Sci. Petrograd; For. Mem. Real. Accad. dei Lincei, Rome, K. Gesell. Wissen. Gött., Kongl. Danske Vidensk. Selskab, Acad. Roy. Méd. de Belgique; Hon. Fellow Roy. Soc. Edin., Roy. Med. Soc. Edin., Hon. Mem. Amer. Physiological Society; Corresp. Mem. Société de Biologie, Paris; For. Mem. Akad. Wetensch., Amsterd.; Silliman Memorial Lecturer Yale University. **Medal: Royal.** 9 *Chadlington Road, Oxford.*	1900-02

Date of Election		Service on Council, &c.
1904	Shipley, Arthur Everett, Sc.D. (Camb.) Hon. D.Sc. (Princeton) F.L.S. F.Z.S. Master of Christ's College, Cambridge. Reader in Zoology, University of Cambridge; Chairman of Council, Marine Biological Assoc. *Christ's College Lodge, Cambridge.*	1909–11
1915	Simpson, George Clarke, D.Sc. (Manch., Syd.) Meteorologist to the Government of India. *Meteorological Office, Simla, India.*	
1907	Smith, Grafton Elliot, M.A. (Camb.) M.D. (Sydney). Professor of Anatomy in the University of Manchester. Formerly Professor of Anatomy in the Government School of Medicine, Cairo. Ord. Osmanieh Com.; Mem. Corr. Soc. Anthrop. Paris; Ehrenmitglied Anthrop. Gesell. Munich; For. Mem. Soc. Anthrop. Rome. **Medal: Royal.** *The University, Manchester; and Moss Side, Maghull, near Liverpool.*	1912–14 V.P. 1913–1
1909	Smith, James Lorrain, M.A. M.D. (Edin.) M.Sc. (Manch.). Professor of Pathology in the University of Edinburgh. 11 *Bruntsfield Crescent, Edinburgh.*	1913–14
1914	Smith, Samuel Walter Johnson, M.A. (Camb.) D.Sc. (Lond.) Assistant Professor of Physics in the Imperial College of Science and Technology, South Kensington, S.W. 12 *Fordhook Avenue, Ealing,* W.	
1901	Smithells, Arthur, B.Sc. (Lond. Vict.) F.I.C. Professor of Chemistry in the University of Leeds. 29 *Moor Road, Far Headingley, Leeds.*	1912–13 1915–
1910	Soddy, Frederick, M.A. (Oxon). Professor of Chemistry in the University of Aberdeen. 14 *Albyn Terrace, Aberdeen.*	
1889	Sollas, William Johnson, Sc.D. (Camb.) LL.D. (Dubl.) Ph.D. (Christiania) D.Sc. (Bristol) Professor of Geology in the University of Oxford; Fellow of University College. Oxford. F.R.S.E. F.G.S. **Medal: Royal.** 48 *Woodstock Road, Oxford; and Athenæum Club.*	1906–07
1900	Spencer, W. Baldwin, C.M.G. B.A. (Oxon) M.A. (Melb.). Professor of Biology in the University of Melbourne; Fellow of Lincoln College, Oxford; Corr. Mem. Z.S. *The University, Melbourne, Victoria.*	
1908	Stapf, Otto, Ph.D. (Vienna) Sec. L.S. Keeper of Herbarium and Library, Royal Botanic Gardens, Kew. Corr. Mem. Deutsch. Bot. Gesell.; Hon. Mem. Bot. Verein Prov. Brandenburg. *Herbarium House, Kew.*	

Date of Election		Service on Council, &c.
1914	Stanton, Thomas Edward, D.Sc. (Vict.) M.Inst.C.E., Superintendent, Engineering Department, National Physical Laboratory, Teddington. 36 *Gloucester Road, Teddington*, S.W.	
1899	Starling, Ernest Henry, M.D. F.R.C.P. Hon. M.D. (Breslau) D.Sc. (Dubl.). Jodrell Professor of Physiology in the University of London. Corr. Mem. K.K. Gesell. d. Aerzte, Vienna; For. Mem. Acad. Roy. Méd. Belg. **Medal: Royal.** 40 *West End Lane, Hampstead*, N.W.	**1905–07** **1915–**
1903	Stead, John Edward, Hon. D.Met. (Sheffield) Hon. D.Sc. (Leeds) F.I.C. F.C.S. Public and Agricultural Analyst for the County Borough of Middlesbrough. Mem. Soc. Public Analysts, Iron and Steel Inst.; Hon. Mem. Amer. Inst. Mining Eng. 11 *Queen's Terrace, Middlesbrough.*	
1896	Stebbing, Rev. Thomas Roscoe Rede, M.A. (Oxon) B.A. (Lond.) F.L.S. F.Z.S., Hon. Fellow Worcester Coll. Oxon, Fellow of King's Coll., London, Hon. Mem. New Zealand Inst. *Ephraim Lodge, The Common, Tunbridge Wells.*	
1893	Stirling, Edward Charles, C.M.G. M.A. M.D. Sc.D. (Camb.) F.R.C.S Professor of Physiology in the University of Adelaide; C.M.Z.S. Consulting Surgeon, Adelaide Hospital; Hon. Director of the South Australian Museum. *The University, Adelaide, South Australia.*	
1902	Stirling, Right Hon. Sir James, Kt. M.A. (Camb.) LL.D. (Aberd.). *Finchcocks, Goudhurst, Kent.*	**1908–10** **V.P.** 1909–
1911	Stoney, George Gerald, B.A.I. (Dubl.) M.Inst.C.E. *Oakley, Heaton Road North, Newcastle-on-Tyne.*	
1903	Strahan, Aubrey, M.A. Sc.D. (Camb.) Hon. LL.D. Toronto. F.G.S. F.R.G.S. Director of the Geological Survey of Great Britain, and of the Museum of Practical Geology, London. 28 *Jermyn Street*, S.W.; 12 *Marloes Road, Kensington*, W.; *and Athenæum Club.*	**1909–**
1905	Strutt, Hon. Robert John, M.A. (Camb.). Professor of Physics in the Imperial College of Science and Technology, South Kensington, S.W.; Late Fellow of Trinity College, Cambridge. 69 *Cadogan Square*, S.W.	
1888	Sudeley, Right Hon. Charles Douglas Richard Hanbury-Tracy, Lord. *Ormeley Lodge, Ham Common, Surrey.*	•
1906	Swinburne, James, M.Inst.C.E. Mem. Inst. Elect. Eng. *Woodhurst, Oxted, Surrey.*	
1915	Swinton, Alan Archibald Campbell, M.Inst.C.E. M.I.E.E. 66 *Victoria Street*, S.W.; *and Athenæum Club.*	
1896	Sydenham, George Sydenham Clarke, Lord, G.C.S.I. G.C.M.G. G.C.I.E. 101 *Onslow Square*, S.W.; *and Athenæum and United Service Clubs.*	
1903	Symington, Johnson, M.D. Professor of Anatomy, Registrar and Member of Senate, Queen's University, Belfast. F.Z.S. F.R.S.E. *Queen's University, Belfast.*	

Date of Election		Service on Council, &c.
1915	Tansley, Arthur George, M.A. (Camb.) F.R.G.S. University Lecturer in Botany, Cambridge. *Botany School, Cambridge; and Grantchester, Cambridge.*	
1898	Taylor, Henry Martyn. Fellow of Trinity College, Cambridge. *The Yews, Queen's Road, Cambridge.*	
1888	Teale, Thomas Pridgin, M.A. F.R.C.S. *North Grange, Headingley, Leeds.*	
1890	Teall, J. J. H., M.A. Sc. D. (Camb.) Hon. D.Sc. (Dubl. Oxon.), Hon. LL.D. (St. Andr.) F.G.S. Lately Director of the Geological Survey of Great Britain, and of the Museum of Practical Geology. 174 *Rosendale Road, Dulwich; and Athenæum Club.*	1899–1901 1912–13 V.P. 1900–01
1880	Thiselton-Dyer, Sir William Turner, K.C.M.G. C.I.E. M.A. (Oxon) Sc.D. (Camb.) D.Sc. (Oxon) B.Sc. (Lond.) Ph.D. LL.D. (Glasg.) F.L.S. Lately Director Royal Botanic Gardens, Kew; lately Botanical Adviser to H.M. Secretary of State for the Colonies; Hon. Student of Christ Church, Oxford; Hon. Prof. Roy. Agric. Coll. Cirencester; lately Fellow Univ. of London; Hon. Fellow, King's Coll., Lond., Roy. Soc. Edin., Bot. Soc. Edin., Roy. Soc., South Africa; Hon. Mem. Roy. Bot. Soc. Lond., Pharm. Soc. Gt. Britain, Camb. Phil. Soc., Lit. Phil. Soc. Manchester, Soc. Nat. Sci. Lit. and Art, Chester, Soc. Néerland. d'Hort. et de Bot., New Zealand Institute, Roy. Soc. N.S.W., Dominica Agric. Soc.; Corresp. Acad. Sci. Philad., Boston Soc. Nat. Hist., Hort. Soc. Berlin and Massachusetts, Soc. Nat. Sci. et Math. de Cherb., Corresp. Accad. Sci. Lett. ed Arti d. Zelanti, Acireale, and Botan. Soc. Copenhagen; Mem. Assoc. Soc. Roy. de Bot. de Belgique, American Phil. Soc.; Mitg. Kais.-Leop.-Carol. Deutsch. Acad. der Naturf. in Halle. *The Ferns, Witcombe, Gloucester; and Athenæum Club.*	1886–88 96–97 V.P. 1896–97
1901	Thomas, M. R. Oldfield, F.Z.S. F.R.G.S. Senior Assistant in the Zoological Department of the British Museum. *Natural History Museum, Cromwell Road,* S.W.; *and* 15 *St. Petersburg Place, Bayswater,* W.	1910–11
1891	Thompson, Silvanus Phillips, B.A. D.Sc. (Lond.) LL.D. (Birm. Bristol). M.D. (Königsberg) Principal and Professor of Physics in the City and Guilds of London Technical College, Finsbury, E.C. F.R.A.S.; Reg. Acad. Sci. Suec. Soc.; Phys. Verein, Francof. ad Mœnum, Amer. Philos. Soc., Soc. Philos. Ebor. Soc. Honor.; F. Phys. Soc., Mem. Inst. Elect. Eng.; *Morland, Chislett Road, West Hampstead,* N.W.; *and Athenæum Club.*	1906–07 1911–13
1897	Thomson, John Millar, LL.D. (Glasg.) F.C.S. F.I.C. Hon. Fellow King's Coll. and Queen's Coll., London. Corr. Mem. Roy. Phil. Soc., Glasgow. 18 *Lansdowne Road, Holland Park,* W.	

Date of Election		Service on Council, &c.
1884	Thomson, Sir Joseph John, Kt. O.M.—**President**—M.A. Sc.D. (Dubl.) D.Sc. (Oxon, Columbia, Johns Hopkins, Vict.) LL.D. (Glasg. Aberd. Princeton, Birm., St. And. Gött.) Ph.D. (Cracow, Christiania), Fellow of Trinity College and Cavendish Professor of Experimental Physics, Cambridge; Professor of Physics at the Royal Institution. F.R.S.E. For. Mem. Inst. France; Hon. Mem. Lit. Phil. Soc. Manch., Roy. Dubl. Soc., R. Accad. Sci. Turin, K. Vetensk.-Soc. Upsala; For. Mem. Amer. Acad. Arts and Sci., Amer. Phil. Soc., R. Acad. Phys. Math. Naples, Phys. Med. Soc. Erlangen, R. Accad. Linc. Rome, K. Pruss. Akad. Wissen. Berlin, Batav. Phil. Soc. Rotterdam, R. Acad. Sci. Netherlands, Acad. Sci. Bologna; Nobel Laureate, Physics, 1906. **Medals: Copley, Royal, Hughes.** *Holmleigh, West Road, Cambridge.*	1889–91 1898–1900 1911–13 1915– **V.P.** 1911–13 **Pres.** 1915–
1893	Thornycroft, Sir John Isaac, Kt. LL.D. M.Inst.C.E. *Eyot Villa, Chiswick Mall, Chiswick.*	1913–15
1908	Thorpe, Jocelyn Field, D.Sc. (Manch.) Ph.D. (Heidel.) F.C.S. F.I.C. Professor of Organic Chemistry in the Imperial College of Science and Technology; Fellow of King's Coll., Lond. Formerly Sorby Research Fellow Roy. Soc. *Imperial College of Science and Technology, South Kensington,* S.W.	
1876	Thorpe, Sir T. Edward, Kt. C.B. D.Sc. (Vict. Leeds) Sc.D. (Dubl.) Ph.D. (Heid.) LL.D. (Glasg.) F.C.S. Formerly Principal of the Government Laboratories; Emeritus Professor of Chemistry, Imperial College of Science and Technology, South Kensington; Fellow of the University of London; Hon. Fellow Roy. Soc. Edin.; Mem. Soc. Chem. Indust.; Soc. Chem. Berol. Socius; Soc. Phil. Glasc. Mem. Corr., Imp. Acad. Sci. Petrograd, Mem. Corr.; Soc. Phil. Leeds, Soc. Lit. Phil. Manc., Soc. Pharm., Soc. Pub. Anal. Soc. Honor.; Soc. Bat. Sci. Harl. Soc. Extr. **Medal: Royal.** *Whinfield, Salcombe, South Devon; and Athenæum Club.*	1890–91 93–95 99–03 **V.P.** 1894–95 **For. Sec.** 1899–1903
1899	Threlfall, Richard, M.A. *Oakhurst, Church Rd., Edgbaston, Birmingham.*	1915–
1880	Tilden, Sir William Augustus, Kt. D.Sc. (Lond.) Sc.D. (Dubl.) D.Sc. (Vict.) LL.D. (Birm.) F.C.S. F.I.C. Formerly Professor of Chemistry and Dean of the Royal College of Science, London; Fellow of the University of London; Corr. Mem. Imp. Acad. Sci. Petrograd; Hon. Mem. Pharm. Soc., Soc. Pub. Anal., Soc. Nat. Bristol, Phil. Soc. Birmingham. Coll. Pharm. Philad.; Professor Emeritus in the Imperial College of Science and Technology, South Kensington. **Medal: Davy.** *The Oaks, Northwood, Middlesex.*	1892-94 1904–06 **V.P.** 1904–06
1891	Tizard, Thomas Henry, Captain R.N. C.B. F.R.G.S. Lately Assistant Hydrographer of the Admiralty. 23 *Geneva Road, Kingston-on-Thames.*	1902–04
1878	Tomes, Charles Sissmore, M.A. (Oxon) LL.D. (Birm.). *Mannington Hall, Aylsham, Norfolk.*	
1889	Tomlinson, Herbert, B.A. (Oxon). *Gilling, Upper Station Road, Bexhill-on-Sea.*	

Date of Election		Service on Council, &c.
1903	Townsend, John S., M.A. (Dubl.). Wykeham Professor of Physics, Oxford. **Medal: Hughes.** 20 *Merton Street, Oxford.*	1914–15
1893	Trail, James William Helenus, A.M. M.D. C.M. (Aberd.) F.L.S. Regius Professor of Botany in the University of Aberdeen. *The University, Aberdeen, N.B.*	
1904	Travers, Morris William, D.Sc. (Lond. Bristol). Fellow Univ. Coll. Lond.; lately Director of the Indian Institute of Science, Bangalore. 17 *Lexham Gardens,* W.	
1883	Trimen, Roland, Hon. M.A. (Oxon) F.L.S. F.Z.S. F.E.S. Hon. Mem. South African Phil. Soc. and Soc. Imp. Amis. Sci. Nat. de Mosc.; lately Curator of the South African Museum.; Corr. Mem. Amer. Entom. Soc., Philad. **Medal: Darwin.** 33 *Croftdown Road, Highgate Road,* N.W.	
1897	Trouton, Frederick Thomas, M.A. Sc.D. (Dubl.) Lately Professor of Physics in the University of London. *The Rydings, Redington Road, Hampstead,* N.W.; *and University College,* W.C.	1907–09
1897	Turner, Herbert Hall, D.Sc. (Oxon, Leeds, Sydney) D.C.L. (Durham) Savilian Professor of Astronomy in the University of Oxford. F.R.A.S., lately Fellow Trinity Coll. Camb.; Fellow New Coll. Oxford; Corresp. Inst. Fr. (Acad. Sci.); Corr. Mem. Soc. Spettros. Ital. *University Observatory, Oxford.*	1901–03
1877	Turner, Sir William, K.C.B. M.B. (Lond.) D.C.L. (Durh. Toronto and Oxf.) LL.D. (Glasg. St. Andr. Aberd. Western Univ. Pennsylvania and Univ. McGill) Sc.D. (Camb. and Dubl.) F.R.C.S. (Lond. Edin.) F.R.S.E.; Vice-Chancellor, Principal and Emeritus Professor of the University of Edinburgh; Hon. Assoc. Ord. Hosp. St. John, Jerusalem; Hon. Prof. Anat. Roy. Soc. Acad.; Hon. Mem. Roy. Irish Acad.; Hon. Fell. Roy. Med. Chir. Soc. London; Hon. Fell. Obst. Soc. Lond. and Edin.; For. Assoc. Anthrop. Soc. Paris; Corr. Mem. Akad. Wiss. Berlin, Soc. Anthrop. Ethnol. and Prehist. Arch. Berlin. Corr. Mem. Soc. Anthrop. Rome, Brussels; Hon. Mem. Imp. Milit. Acad. Med. Petrograd; Knt. Pruss. order "*Povr le Mérite.*" 6 *Eton Terrace, Edinburgh; and Athenæum Club.*	1890–91
1899	Tutton, Alfred Edwin Howard, M.A. (Oxon) D.Sc. (Oxon) F.C.S. A.R.C.S. *Duart, Yelverton, S. Devon.*	
1871	Tylor, Sir Edward Burnett, Kt. D.C.L. (Oxon) LL.D. (St. And. Aberd. and McGill) Assoc. Acad. Reg. Belg. Formerly Professor of Anthropology in the University of Oxford. *Linden, Wellington, Somerset.*	1899–1900
1886	Unwin, W. Cawthorne, B.Sc. LL.D. (Edin.) Mem. Inst. C.E.; Hon. Mem. Inst. M.E.; Mem. Amer. Phil. Soc.; Hon. Mem. Amer. Soc. Mech. Eng.; Emeritus Professor of Engineering at the Central Technical College of the City and Guilds of London Institute. *Palace Gate Mansions,* 29 *Palace Gate, Kensington,* W.	1893–94

Date of Election		Service on Council, &c.
1894	Veley, Victor Herbert, M.A. D.Sc. 8 *Marlborough Place, St. John's Wood*, N.W.	
1883	Venn, John, Sc.D., Fellow of Caius College, Cambridge. *Vicarsbrook, Chaucer Road, Cambridge.*	
1885	Vines, Sydney Howard, M.A. (Oxon) D.Sc. (Camb. and Lond.) Sherardian Professor of Botany in the University of Oxford; Fellow of Magdalen College, Oxford; Hon. Fellow of Christ's College, Cambridge; Fellow Univ. London; F.L.S.; Hon. Mem. Manc. Lit. Phil. Soc. and Roy. Phys. Soc. Edin.; Corr. Mem. Soc. Nat. Sci. et Math. de Cherb., Soc. Roy. Bot. de Belg., and Soc. Nat. Hist. Bost. *Headington Hill, Oxford.*	1890–92
1904	Wager, Harold W. T., F.L.S. H.M. Inspector of Schools, Secondary Branch, Board of Education; sometime Lecturer on Botany in the University, Leeds. *Hendre, Horsforth Lane, Far Headingley, Leeds.*	1910–12
1913	Walker, George Walker, M.A. A.R.C.S. Formerly Fellow of Trinity College, Cambridge. *Heath Cottage, Boar's Hill, near Oxford.*	
1904	Walker, Sir Gilbert Thomas, Kt., C.S.I. Sc.D. M.A. F.R.A.S. Director General of Observatories, India; lately Lecturer and Fellow of Trinity College, Cambridge. *Red Roof, Simla, India.*	
1900	Walker, James, D.Sc. (Edin.) Ph.D. (Leip.), LL.D. (St. Andr.). Professor of Chemistry in the University of Edinburgh. 5 *Wester Coates Road, Edinburgh.*	
1892	Waller, Augustus Désiré, M.D. (Aberd.) Hon. LL.D. (Edin.) Director of the Physiological Laboratory, University of London. *Weston Lodge*, 32 *Grove End Road*, N.W.	1906–08
1887	Walsingham, Thomas de Grey, Lord, M.A. LL.D. High Steward of the University of Cambridge; Trust. Brit. Mus.; F.L.S. F.Z.S. F.E.S.; Mem. Soc. Ent. de France, Ent. Ver. zu Berlin, Nederlands Ent. Ver., Soc. Ent. de Russie, Linn. Soc. N.S.W. *Merton Hall, Thetford, Norfolk.*	1896–97
1884	Warren, Sir Charles, General, R.E. G.C.M.G. K.C.B. *The Oaks, Westbere, Canterbury; and Athenæum Club*, S.W.	
1901	Watson, William, D.Sc. A.R.C.S. (Lond.) Assistant Professor of Physics in the Imperial College of Science and Technology, London. 7 *Upper Cheyne Row, Chelsea*, S.W.	
1900	Watts, Sir Philip, K.C.B. VICE-PRESIDENT. Lately Director of Naval Construction. Order of the Rising Sun (Second Class). 10 *Chelsea Embankment*, S.W.	1912–13 1915– V.P. 1915–
1904	Watts, William Whitehead, LL.D. (St. Andr.) Sc.D. (Camb.) M.Sc. (Birm.) Professor of Geology in the Imperial College of Science and Technology. F.G.S. Hon. Fellow of Sidney Sussex Coll., Camb.; Hon. Mem. Geol. Soc., Liverp. *Imperial College of Science and Technology, South Kensington*, S.W.; *and Hillside, Langley Park, Sutton, Surrey.*	1912–14

Date of Election		Service on Council, &c.
1901	Whetham, William Cecil Dampier, M.A. Lecturer in Physics, Fellow and Tutor of Trinity College, Cambridge. *Upwater Lodge, Cambridge*; *Cadhay House, Ottery St. Mary, Devon.*	
1887	Whitaker, William, B.A. (Lond.) F.G.S. Assoc. Inst. C.E. F.R. San. Inst.; Corr. Acad. Nat. Sci. Philad.; Hon. Mem. Geol. Assoc. Lond., Phil. Soc. York, Geol. Socs. Liverpool, Manchester and Yorkshire, Soc. Belg. de Géol. 3 *Campden Road, Croydon.*	1907–09
1903	Whitehead, Alfred North, Sc.D. (Camb.) Professor of Applied Mathematics in the Imperial College of Science and Technology, South Kensington, S.W.; Fellow and sometime Lecturer in Mathematics, Trinity College, Cambridge. 12 *Elm Park Gardens, Chelsea,* S.W.	1913–15
1905	Whittaker, Edmund Taylor, M.A. (Camb.) Hon. Sc.D. (Dubl.). Professor of Mathematics in the University of Edinburgh. 35 *George Square, Edinburgh.*	1911–12
1886	Wilde, Henry, D.Sc. D.C.L. (Oxon). Mem. Lit. Phil. Soc. Manch., Hon. Mem. Inst. Elect. Eng. *The Hurst, Alderley Edge, Cheshire.*	
1902	Willey, Arthur, D.Sc. (Lond.) Hon. M.A. (Camb.) Professor of Zoology in McGill University, Montreal; Corresp. Mem. New York Acad. Sci., Fellow of University College, London. *Zoological Department, McGill University, Montreal, Canada.*	
1879	*Williamson, Benjamin, D.Sc. D.C.L. (Oxon) M.R.I.A. Senior Fellow of Trinity College, Dublin. *Trinity College, Dublin.*	
1900	Wilson, Charles Thomson Rees, M.A. (Camb.) B.Sc. (Vict.) Fellow of Sidney Sussex College, Cambridge; University Lecturer in Physics, Cambridge. **Medal: Hughes.** 14 *Cranmer Road, Cambridge.*	1913–15
1906	Wilson, Harold A., M.A. (Camb.) D.Sc. (Lond.) M.Sc. (Vict.). Professor of Physics, Rice Institute, Houston, Texas; formerly Fellow of Trinity College, Cambridge; lately Macdonald Professor of Physics, McGill University, Montreal; formerly Professor of Physics in King's College, London. *Rice Institute, Houston, Texas, U.S.A.*	
1909	Wilson, James Thomas, M.B. (Edin.) Challis Professor of Anatomy in the University of Sydney. Hon. Fellow Roy. Soc. S. Aust.; Corr. Mem. Gesell. für Ethnol. Anthrop. Berlin. *The University, Sydney, N.S.W.*	
1899	Windle, Sir Bertram Coghill Alan, Kt. M.A. M.D. Sc.D. (Dubl.) Hon. LL.D. (Roy. Univ. Ireland, Birm.), M.Sc. (Birm.) President of University College, Cork; F.S.A. F.R.Soc. Antiq. Ireland, M.R.I.A. *President's House, University College, Cork.*	
1895	Wolfe Barry, Sir John, K.C.B. LL.D. Mem. Inst. C.E. *Delahay House,* 15 *Chelsea Embankment,* S.W.	1902–
1901	Woodward, Arthur Smith, LL.D. (St. Andr. Glasg.) F.L.S. Pres.G.S. F.Z.S. F.R.G.S. Keeper of the Department of Geology, British Museum (Natural History). 4 *Scarsdale Villas, Kensington,* W.	1913–15

* Deceased, January, 1916.

Date of Election | Service on Council, &c.

1873 Woodward, Henry, LL.D. (St. And.) F.G.S. V.P.Z.S. F.R.M.S. Pres. Palæont. Soc. Mem. Malacol. Soc. Lond. Acad. Sci. Nov. Ebor. Soc. Phil. Amer. Philad. Soc.; Soc. Phil. Ebor., Assoc. Geol. Lond., Socc. Geol. Edin., Glasc., Liverp. et Nordov. Soc. Honor.; Socc. Géol. Belg., Imp. Nat. Hist. Mosq., Hist. Nat. Montreal et Malacol. Belg. Corresp.; lately Keeper of the Department of Geology, British Museum (Natural History). 13 *Arundel Gardens, Notting Hill*, W.

1893 Worthington, Arthur Mason, C.B. M.A. F.R.A.S. Late Professor of Physics at the Royal Naval College, Greenwich. Formerly Head-master and Professor of Physics, Royal Naval Engineering College, Devonport. 5 *Louisa Terrace, Exmouth.*

1906 Wright, Sir Almroth Edward, M.D. (Dubl.) Hon. D.Sc. (Dubl.). Director of the Department for Therapeutic Inoculation, St. Mary's Hospital, London; Professor of Experimental Pathology, University of London; formerly Professor of Pathology, Army Medical School, Netley. Hon. F.R.C.S. (Ireland). Hon. Fell. Med. Soc. Sweden; Lecomte Prize, Acad. Sci. Paris, 1915. 6 *Park Crescent*, W.; *and St. Mary's Hospital*, W.

1896 Wynne, William Palmer, D.Sc. (Lond.) F.I.C. Assoc. R.C.S. Firth Professor of Chemistry and Dean of the Faculty of Pure Science in the University of Sheffield. 17 *Taptonville Road, Sheffield; and The University, Sheffield.*

1893 Young, Sydney, D.Sc. (Lond.) Hon. Sc.D. (Dubl. Bristol) Hon. D.Sc. (Calcutta) B.Sc. (Manch.) Professor of Chemistry in Trinity College, Dublin. F.C.S. F.I.C. M.R.I.A. 12 *Raglan Road, Dublin.*

1907 Young, William Henry, M.A. Sc.D. (Camb.), Professor of the Philosophy and History of Mathematics in the University of Liverpool. Hon. Dr. ès Sc. Math. (Genève). *Epinettes* 22, *Lausanne, Switzerland.*

FOREIGN MEMBERS.

Date of Election		Medal.
1910.	Arrhenius, Svante August. Nobel Laureate, Physics, 1903. *Nobel Institute, Stockholm*	**Davy.**
1911.	Backlund, Jöns Oskar. *Observatoire Central Nicolas, Poulkovo*	
1885.	Baeyer, Adolf von. Nobel Laureate, Chemistry, 1905. *Universität, Munich*	**Davy.**
1913.	Barrois, Charles. 37 *Rue Pascal, Lille*	
1902.	Brøgger, Waldemar Christofer. *K. Frederiks Universitet, Christiania*	
1913.	Chatelier, Henry Louis le. *Rue Notre Dame des Champs*, 75 *Paris*	
1889.	Chauveau, Jean Baptiste Auguste. *Rue du Cloître-Notre-Dame*, 4, *Paris*	
1902.	Darboux, Jean Gaston. *Secrétariat de l'Institut, Paris*	
1905.	De Vries, Hugo. *Universiteit, Amsterdam*	**Darwin.**
1899.	Fischer, Emil. Nobel Laureate, Chemistry, 1902. *Universität, Berlin*	**Davy.**
1911.	Groth, Paul Heinrich Ritter von. *Universität, Munich*	
1909.	Hale, George Ellery. *Astrophysical Observatory, Mount Wilson, California, U.S.A.*	
1896.	Heim, Albert. *Hochschule, Zürich*	
1908.	Helmert, Friedrich Robert. *Universität, Berlin*	
1902.	Hering, Ewald. *Universität, Leipzig*	
1911.	Kayser, Heinrich. *Universität, Bonn*	
1908.	Kitasato, S. *Imperial Pathological Laboratory, Tokio*	
1885.	Klein, Felix. *Wilhelm Weber Strasse* 3, *Göttingen*	**Copley.**
1911.	Le Bel, Joseph Achille. 250 *Rue Saint Jacques, Paris*	**Davy.**
1896.	Lippmann, Gabriel. Nobel Laureate, Physics, 1908. *Faculté des Sciences à la Sorbonne, Paris*	
1905.	Lorentz, Hendrik Antoon. Nobel Laureate, Physics, 1902. *Zijlweg* 76, *Haarlem*	**Rumford.**
1895.	Metchnikoff Elias. Nobel Laureate, Medicine, 1908. *Institut Pasteur, Paris*	**Copley.**
1902.	Michelson, Albert Abraham. Nobel Laureate, Physics, 1907. *University, Chicago, U.S.A.*	**Copley.**
1896.	Mittag-Leffler, Gösta. *Djursholm, Sweden*	
1907.	*Pavlov, Ivan Petrovitch. Nobel Laureate, Physiology and Medicine, 1904. *Wedenskaja* 4, *Petrograd*	**Copley.**
1897.	Pfeffer, Wilhelm. *Universität, Leipzig*	
1909.	Picard, Charles Emile, 4 *Rue Joseph Bara, Paris*	
1907.	Pickering, Edward Charles. *Harvard College Observatory, Cambridge, Mass., U.S.A.*	
1879.	Quincke, Georg Hermann. *Bergstrasse* 41, *Heidelberg*	

* Deceased, January, 1916.

Date of Election		Medal.
1909.	Ramón y Cajal, Santiago. Nobel Laureate, Medicine, 1906. *Universidad, Madrid*	
1907.	Retzius, Magnus Gustaf. *Högskolan, Stockholm*	
1907.	Righi, Augusto. *Regia Università, Bologna, Italy*	**Hughes.**
1913.	Roux, Émile. *Institut Pasteur, Paris*	
1913.	Schwendener, Simon. *Matthäikirchstrasse* 28, *Berlin*	
1911.	Timiriazeff, Clement A. *Université, Moscow*	
1913.	Voigt, Woldemar. *Universität, Göttingen*	
1910.	Volterra, Vito. *Regia Università, Rome*	

FELLOWS DECEASED DURING THE YEAR 1915.

On the Home List.

Adams, William Grylls.
Alverstone, Lord.
Assheton, Richard.
Bastian, Henry Charlton.
Church, Sir Arthur Herbert.
Crofton, Morgan William.
Cunningham, David Douglas.
Debus, Heinrich.
Geikie, James.
Gowers, Sir William Richard.
Lea, Arthur Sheridan.
Lydekker, Richard.
Meldola, Raphael.
Minchin, Edward Alfred.
Müller, Hugo.
Nares, Sir George Strong.
Neville, Francis Henry.
Noble, Sir Andrew.
Roscoe, Sir Henry Enfield.
Rücker, Sir Arthur William.
Sell, William James.
Tanner, Henry William Lloyd.
Tennant, Lieut.-Gen. James Francis.

On the Foreign List.

Amagat, Émile Hilaire.
Auwers, Georg Friedrich Julius Arthur.
Ehrlich, Paul.
Solms-Laubach, Graf Hermann zu.

FELLOWS ELECTED DURING THE YEAR 1915.

Andrewes, Frederick William.
Conway, Arthur William.
Doncaster, Leonard.
Evershed, John.
Fletcher, Walter Morley.
Green, Arthur George.
Hayden, Henry Hubert.
Mackenzie, James.
McLennan, John Cunningham.
Masterman, Arthur Thomas.
Morgan, Gilbert Thomas.
Myers, Charles Samuel.
Simpson, George Clarke.
Swinton, Alan A. Campbell.
Tansley, Arthur George.

CHANGE OF TITLE.

Rothschild, Hon. Lionel Walter *to* Rothschild, Lord.

COMMITTEES, 1916.

NOTE.—*The Chairmen of Committees named below are appointed to act as Chairmen in the absence of the President, who by Statute presides over all Committees which he attends.*

The Treasurer is ex-officio *member, and one of the Secretaries shall be nominated member, of each Committee excepting the Scientific Relief Committee, the Sectional Committees, and any Committee composed of representatives of the Royal Society and other Bodies jointly. Either the Treasurer or one of the Secretaries shall be a member of any Committee composed of representatives of the Royal Society and other Bodies jointly.*

Each Committee, excepting those specified in the previous paragraph, has power to suggest to the Council names of additional members for appointment on such Committee: the names not being necessarily those of Fellows of the Society.

STANDING AND OCCASIONAL COMMITTEES.

CATALOGUE OF SCIENTIFIC PAPERS COMMITTEE.

Chairman—Prof. Silvanus Thompson.

Prof. H. E. Armstrong, Prof. W. Bateson, Sir John Bradford, Mr. H. T. Brown, Sir Frank Dyson, Prof. J. B. Farmer, Sir Archibald Geikie, Mr. W. B. Hardy, Dr. E. Klein, Sir Joseph Larmor, Dr. H. McLeod, Dr. C. J. Martin, Dr. P. C. Mitchell, Dr. D. H. Scott, Prof. A. C. Seward, Prof. A. Schuster, Dr. A. Strahan, Prof. Silvanus Thompson, Sir Edward Thorpe, and Sir William Tilden.

"CHALLENGER" REPORTS COMMITTEE.

Chairman—Sir Ray Lankester.

Sir John Bradford, Mr. J. Y. Buchanan, Prof. J. S. Gardiner, Prof. W. A. Herdman, Sir Ray Lankester, Dr. D. H. Scott, and Sir William Thiselton-Dyer.

FINANCE COMMITTEE.

Chairman—The Treasurer.

The Treasurer, Mr. G. T. Beilby, Mr. H. T. Brown, Sir Robert Hadfield, Sir Joseph Larmor, Dr. P. C. Mitchell, Sir Charles Parsons, Sir James Stirling, Sir Edward Thorpe, and Dr. A. N. Whitehead.

GASSIOT COMMITTEE.

Chairman—Major H. G. Lyons.

The Astronomer Royal, Dr. C. Chree, Mr. W. H. Dines, Sir Archibald Geikie, Dr. R. T. Glazebrook, Sir Joseph Larmor, Major H. G. Lyons, Prof. H. F. Newall, Prof. A. Schuster, Sir Napier Shaw, Mr. G. W. Walker, Prof. W. Watson, and Mr. C. T. R. Wilson.

Glass Workers' Cataract Committee.

Chairman—Sir William Abney.

Sir William Abney, Sir Clifford Allbutt, Dr. H. K. Anderson, Sir John Bradford, Sir William Crookes, Mr. W. B. Hardy, and Dr. A. D. Waller, with Mr. J. H. Parsons.

Indian Government Advisory Committee.

Chairman—Lt.-Col. Sir David Prain.

Lt.-Col. A. W. Alcock, Prof. W. Bateson, Prof. R. H. Biffen, Sir John Bradford, Mr. H. T. Brown, Surg.-Gen. Sir David Bruce, Prof. W. R. Dunstan, Mr. J. S. Gamble, Sir Joseph Larmor, Sir Alexander Pedler, Lt.-Col. Sir David Prain, Sir William Schlich, Mr. A. E. Shipley, and Sir William Thiselton-Dyer, with Sir John McFadyean.

International Association of Academies Committee.

Chairman—Sir Joseph Larmor.

Prof. H. E. Armstrong, Sir John Bradford, Prof. A. R. Forsyth, Mr. W. B. Hardy, Sir Ray Lankester, Sir Joseph Larmor, Sir Norman Lockyer, Lt.-Col. Sir David Prain, Sir Ronald Ross, Dr. D. H. Scott, Prof. A. Schuster, Sir Napier Shaw, Prof. C. S. Sherrington, Prof. Silvanus Thompson, Prof. H. H. Turner, and Dr. A. D. Waller.

International Catalogue Committee.

Chairman—Prof. H. E. Armstrong.

Prof. H. E. Armstrong, Sir John Bradford, Mr. H. T. Brown, Sir William Christie, Prof. J. B. Farmer, Prof. A. R. Forsyth, Prof. W. D. Halliburton, Mr. W. B. Hardy, Prof. C. Lapworth, Sir Norman Lockyer, Prof. A. E. H. Love, Dr. H. McLeod, Major P. A. MacMahon, Sir Henry Miers, Dr. P. C. Mitchell, Lt.-Col. Sir David Prain, Dr. D. H. Scott, Prof. A. Schuster, Mr. D. Sharp, Dr. J. J. H. Teall, Mr. Oldfield Thomas, and Sir Edward Thorpe, with Mr. B. Daydon Jackson.

Joint Permanent Eclipse Committee.*

Chairman—The Astronomer Royal.

(On the part of the Royal Society.)

Sir William Abney, The Astronomer Royal, Prof. H. L. Callendar, Sir William Christie, Admiral Sir A. M. Field, Sir Joseph Larmor, Sir Norman Lockyer, Major P. A. MacMahon, Prof. A. Schuster, and Sir Edward Thorpe.

* The following are the Representatives of the Royal Astronomical Society on this Committee:—Rev. A. L. Cortie, Dr. A. C. D. Crommelin, Prof. A. S. Eddington, Prof. A. Fowler (*Secretary*), Major E. H. Hills, Mr. F. K. McClean, Mr. E. W. Maunder, Prof. H. F. Newall, Mr. F. J. M. Stratton, Prof. H. H. Turner.

Library Committee.

Chairman—Prof. G. Carey Foster.

Prof. T. G. Bonney, Sir John Bradford, Prof. J. N. Collie, Mr. W. Duddell, Prof. J. B. Farmer, Prof. G. Carey Foster, Sir George Greenhill, Prof. W. D. Halliburton, Dr. S. F. Harmer, Dr. A. S. Woodward, Mr. G. B. Mathews, Dr. H. McLeod, Prof. A. Schuster, and Prof. Silvanus Thompson.

Magnetic Survey Committee.*

The President, the Treasurer, Prof. A. Schuster, Major H. G. Lyons, Sir Edward Thorpe, and Prof. W. Watson.

Observatories Committee.

(Appointed for ensuring the continued efficiency of the Indian Observatories.)

Chairman—The Astronomer Royal.

The Astronomer Royal, the President of the Royal Astronomical Society, Sir William Abney, Dr. C. Chree, Sir William Christie, Sir Joseph Larmor, Prof. H. F. Newall, Prof. A. Schuster, Sir Napier Shaw, and Prof. H. H. Turner.

Scientific Relief Committee.

Chairman—Sir Clifford Allbutt.

Sir Joseph Larmor, Sir Thomas Barlow, Prof. E. B. Poulton, Lt.-Col. Sir David Prain, Prof. M. J. M. Hill, Dr. R. Messel, Sir Clifford Allbutt, Prof. A. E. H. Love, Prof. E. B. Elliott and Prof. Silvanus Thompson (*in order of appointment*).

(*Two Members, in order of appointment, retire annually.*)

Seismology Committee.

Chairman—Sir Archibald Geikie.

Mr. C. V. Boys, Mr. Horace Darwin, Sir Alfred Ewing, Sir Archibald Geikie, Mr. E. Hull, Prof. A. E. H. Love, Mr. R. D. Oldham, Prof. J. Perry, Dr. J. J. H. Teall, Prof. H. H. Turner, and Mr. G. W. Walker.

Soirée Committee.

Chairman—Lt.-Col. Sir David Prain.

Prof. A. W. Crossley, Prof. J. B. Farmer, Mr. W. B. Hardy, Major H. G. Lyons, Prof. E. W. MacBride, Mr. A. Mallock, Dr. P. C.

* The following are the representatives of the British Association on the Committee :—The Astronomer Royal, Dr. C. Chree, Dr. R. T. Glazebrook, and Prof. H. H. Turner. The Ordnance Survey Department is represented by the Director-General and Captain E. O. Henrici.

Mitchell, Lt.-Col. Sir David Prain, Prof. A. Schuster, Hon. R. J. Strutt, and Dr. A. S. Woodward.

(*Three Members, to be determined by longest service or least attendance, · retire annually.*)

SOLAR RESEARCH COMMITTEE.

Chairman—Prof. H. F. Newall.

The Astronomer Royal, Sir William Christie, Prof. A. Fowler, Sir Joseph Larmor, Sir Norman Lockyer, Prof. H. F. Newall, Prof. A. Schuster, and Prof. H. H. Turner.

SORBY RESEARCH FUND COMMITTEE.

Chairman—The President.

The President, Hon. Sir Charles Parsons, Dr. D. H. Scott, and Sir William Tilden, on the part of the Royal Society; and Prof. W. M. Hicks, Prof. W. P. Wynne, and the Vice-Chancellor of the University, on the part of the University of Sheffield.

SWINE FEVER COMMITTEE.

Chairman—Prof. E. H. Starling.

Sir Clifford Allbutt, Sir John Bradford. Surg.-Gen. Sir David Bruce, Dr. W. Bulloch, The Duke of Devonshire, Dr. A. E. Garrod, Mr. W. B. Hardy, Dr. F. G. Hopkins, Lt.-Col. Sir William Leishman, Dr. C. J. Martin, Sir William Osler, Sir William Power, Mr. A. E. Shipley, Prof. Lorrain Smith, and Prof. E. H. Starling.

TROPICAL DISEASES COMMITTEE.

Chairman—Sir Archibald Geikie.

Lt.-Col. A. W. Alcock, Sir Clifford Allbutt, Sir John Bradford, Surg.-Gen. Sir David Bruce, Sir Archibald Geikie, Dr. A. Harden, Mr. W. B. Hardy, Sir John Kirk, Sir Ray Lankester, Lt.-Col. Sir William Leishman, Dr. C. J. Martin, Dr. F. W. Mott, Prof. G. H. F. Nuttall, Mr. H. G. Plimmer, Sir Ronald Ross, and Mr. A. E. Shipley, with Dr. A. G. Bagshawe, Lt.-Col. Sir R. Havelock Charles, Sir Charles Lucas, Sir John McFadyean, Dr. Guy Marshall, Dr. R. U. Moffat, Sir Stewart Stockman, and a representative of the Colonial Office.

TYNDALL MINING BEQUEST COMMITTEE.

Chairman—Sir Archibald Geikie.

Prof. H. B. Dixon, Sir Archibald Geikie, Dr. E. H. Griffiths, Dr. J. S. Haldane, and Dr. A. Strahan, with the Secretary of the Institution of Mining Engineers, Sir Henry Cunynghame, Prof. John Cadman, and Sir William Garforth.

SECTIONAL COMMITTEES, 1916.

1. Mathematics Committee :—

(Two to retire each year.)

Chairman—Dr. A. N. Whitehead.

	To serve.	
Prof. A. E. H. Love....	1 year.	Retires Dec., 1916.
Dr. A. N. Whitehead ..	1 ,,	,, ,, ,,
Prof. A. R. Forsyth....	2 years.	,, ,, 1917.
Mr. G. H. Hardy......	2 ,,	,, ,, ,,
Sir Joseph Larmor	3 ,,	,, ,, 1918.
Prof. H. H. Turner....	3 ,,	,, ,, ,,

2. Physics and Chemistry Committee :—

(Four to retire each year.)

Chairman—Prof. H. L. Callendar.

	To serve.	
Prof. F. G. Donnan....	1 year.	Retires Dec., 1916.
Prof. A. Fowler	1 ,,	,, ,, ,,
Prof. S. P. Thompson..	1 ,,	,, ,, ,,
Prof. W. P. Wynne....	1 ,,	,, ,, ,,
Prof. B. Hopkinson....	2 years.	,, ,, 1917.
Prof. A. Liversidge	2 ,,	,, ,, ,,
Dr. G. F. Searle	2 ,,	,, ,, ,,
Mr. C. T. R. Wilson....	2 ,,	,, ,, ,,
Prof. H. L. Callendar..	3 ,,	,, ,, 1918.
Dr. J. A. Harker......	3 ,,	,, ,, ,,
Prof. O. W. Richardson..	3 ,,	,, ,, ,,
Dr. A. Scott..........	3 ,,	,, ,, ,,

3. Geology Committee :—

(Three to retire each year.)

Chairman—Mr. G. W. Lamplugh.

	To serve.	
Mr. G. W. Lamplugh ..	1 year.	Retires Dec., 1916.
Dr. J. E. Marr........	1 ,,	,, ,, ,,
Dr. A. S. Woodward ..	1 ,,	,, ,, ,,
Sir Thomas Holland....	2 years.	,, ,, 1917.
Dr. R. Kidston........	2 ,,	,, ,, ,,
Mr. R. D. Oldham	2 ,,	,, ,, ,,
Dr. J. S. Flett........	3 ,,	,, ,, 1918.
Prof. J. W. Gregory...	3 ,,	,, ,, ,,
Sir Henry Miers	3 ,,	,, ,, ,,

4. Botany Committee :—

(Three to retire each year.)

Chairman—Prof. J. B. Farmer.

	To serve.			
Prof. F. Keeble	1 year.	Retires	Dec.,	1916.
Mr. H. N. Ridley	1 "	"	"	"
Mr. H. Wager..........	1 "	"	"	"
Prof. I. B. Balfour......	2 years.	"	"	1917.
Dr. F. F. Blackman	2 "	"	"	"
Prof. J. B. Farmer......	2 "	"	"	"
Prof. H. H. Dixon......	3 "	"	"	1918.
Sir David Prain........	3 "	"	"	"
Mr. C. Reid...........	3 "	"	"	"

5. Zoology Committee :—

(Three to retire each year.)

Chairman—Prof. J. S. Gardiner.

	To serve.			
Prof. J. P. Hill	1 year.	Retires	Dec.,	1916.
Prof. G. H. F. Nuttall...	1 "	"	"	"
Prof. R. Newstead......	1 "	"	"	"
Dr. C. W. Andrews	2 years.	"	"	1917.
Prof. J. S. Gardiner	2 "	"	"	"
Prof. S. J. Hickson	2 "	"	"	"
Prof. W. Bateson.......	3 "	"	"	1918.
Prof. J. G. Kerr........	3 "	"	"	"
Dr. W. H. R. Rivers....	3 "	"	"	"

6. Physiology Committee :—

(Four to retire each year.)

Chairman—Prof. W. M. Bayliss.

	To serve.			
Prof. W. M. Bayliss	1 year.	Retires	Dec.,	1916.
Prof. V. H. Blackman ..	1 "	"	"	"
Dr. J. S. Haldane	1 "	"	"	"
Prof. R. Muir..........	1 "	"	"	"
Mr. J. Barcroft	2 years.	"	"	1917.
Dr. A. Harden	2 "	"	"	"
Prof. J. B. Leathes	2 "	"	"	"
Prof. C. S. Sherrington..	2 "	"	"	"
Dr. H. H. Dale	3 "	"	"	1918.
Dr. T. R. Elliott	3 "	"	"	"
Dr. H. Head	3 "	"	"	"
Mr. L. Hill............	3 "	"	"	"

GOVERNMENT GRANT BOARDS, 1916.

BOARD A.

(Mathematics, Mathematical Physics, Crystallography, and Mathematical Astronomy.)

Chairman—Prof. A. E. H. Love.

	To serve until Dec. 31.
Mr. G. H. Hardy, Prof. A. E. H. Love	1916
Prof. T. J. I'A. Bromwich, Prof. A. C. Dixon..	1917
Major P. A. MacMahon, Sir Henry Miers......	1918
Prof. A. S. Eddington, Mr. J. H. Jeans........	1919

BOARD B.

(Experimental Physics, Observational Astronomy, and Meteorology.)

Chairman—Prof. Silvanus Thompson.

Prof. S. P. Thompson, Mr. C. T. R. Wilson....	1916
Prof. G. C. Barkla, Major E. H. Hills	1917
Mr. W. Duddell, Hon. R. J. Strutt	1918
Prof. W. H. Bragg, Dr. E. H. Griffiths........	1919

BOARD C.

(Chemistry and Metallurgy.)

Chairman—Prof. W. H. Perkin.

Prof. W. Gowland, Prof. James Walker	1916
Prof. H. B. Baker, Sir Robert Hadfield........	1917
Prof. W. H. Perkin, Prof. S. Young	1918
Dr. H. J. H. Fenton, Prof. P. F. Frankland....	1919

Board D.

(Geology, Palæontology, Mineralogy, and Geography.)

Chairman—Mr. G. W. Lamplugh.

	To serve until Dec. 31.
Mr. G. W. Lamplugh, Prof. W. J. Sollas	1916
Dr. F. A. Bather, Mr. R. D. Oldham..........	1917
Mr. A. R. Hinks, Prof. C. Lapworth	1918
Sir Thomas Holland, Dr. J. E. Marr	1919

Board E.

(Botany.)

Chairman—Prof. F. W. Oliver.

Mr. A. D. Hall, Prof. F. W. Oliver	1916
Dr. A. B. Rendle, Prof. S. Vines	1917
Prof. A. C. Seward, Prof. J. W. H. Trail	1918
Prof. R. H. Biffen, Prof. F. E. Weiss..........	1919

Board F.

(Zoology and Comparative Anatomy.)

Chairman—Prof. G. Elliot Smith.

Prof. R. C. Punnett, Prof. G. Elliot Smith	1916
Mr. E. S. Goodrich, Dr. P. C. Mitchell	1917
Prof. W. A. Herdman, Dr. A. S. Woodward....	1918
Prof. A. Dendy, Prof. J. C. Ewart	1919

Board G.

(Animal Physiology and Medical Subjects.)

Chairman—Prof. E. H. Starling.

Sir Ronald Ross, Prof. C. S. Sherrington......	1916
Dr. W. Bulloch, Prof. E. H. Starling	1917
Prof. J. N. Langley, Mr. H. G. Plimmer.......	1918
Prof. J. G. Adami, Prof. R. Muir	1919

STATUTES OF THE ROYAL SOCIETY.

CONTENTS.

CHAPTER I.

Of the Election and Admission of Fellows.

Certificate of Candidature.

1. Every Candidate shall be proposed and recommended by a certificate in writing signed by six or more Fellows, of whom three at least shall certify their recommendation from personal knowledge, but either of the two Secretaries shall have power to sign a certificate on behalf of a Fellow, not resident in the United Kingdom, on receiving a written request from him. The certificate shall specify the name, rank, profession, qualifications, and usual place of residence of the Candidate; and being delivered to the Assistant Secretary, shall be registered, with the date of delivery, in a book to be kept for the purpose, and read at the next ordinary meeting; and, unless otherwise ordered, shall be suspended in some convenient place in the apartments of the Society until the day of election.

2. Of the Fellows who sign a certificate from personal knowledge of the Candidate, two shall undertake the responsibility of acting as

proposer and seconder respectively, in so far that all communications on the subject of the candidature shall take place with one of these. The proposer shall be responsible for informing the Candidate of the Obligation to be subscribed and of the payments to be made to the Society before he can be admitted a Fellow.

Names of Candidates to be announced in January.

3. At the first Ordinary Meeting of the Society in January, the names of all Candidates proposed prior to the first day of that month, and subsequently to the last announcement of the names of Candidates, and also the names of those Candidates whose certificates have been resuspended as hereinafter provided, shall be announced by one of the Secretaries from a list arranged in alphabetical order, without reference to the dates of the certificates of the Candidates; and these certificates shall remain suspended until the day of Election.

List of Candidates to be printed.

4. Not later than the first week in February, a list shall be printed, containing the names of all the Candidates so announced, arranged in alphabetical order, together with the names of the Fellows by whom each Candidate is proposed and recommended; and a copy of such list shall immediately thereafter be sent to every Fellow.

Selection by Council.

5. The Council shall select by ballot from such printed list of Candidates a number not exceeding fifteen, to be recommended to the Society for Election; but no such selection by the Council shall be valid unless eleven Members at least be present and vote, a majority deciding, or in the event of equality the President having a second or casting vote.

Selected Names communicated to Fellows.

6. At the first Ordinary Meeting of the Society, after the Meeting of the Council at which the selection has been made, the President shall read from the Chair the names of the Candidates whom the Council have selected as most eligible, arranged in alphabetical order; and after such Meeting, a circular letter shall be forthwith sent to every Fellow, naming the day and hour of Election, and enclosing a printed list of the selected Candidates, with space for such alterations as any Fellow may determine to make in pursuance of Statute 8.

Date of Election.

7. The election of Fellows not included in the classes referred to in Statutes 11 and 12 of this Chapter, shall take place on the first Thursday of May; unless the Council shall alter the day of Election to any other day in the month of May, in which case due notice of such alteration shall be given to every Fellow.

Election Procedure.

8. On the day of Election two Scrutators shall be nominated by the President, with the approbation of the Society, to assist the Secretaries in examining the lists; and each Fellow present and

voting, shall deliver to one of the Secretaries or Scrutators one of the printed lists mentioned in Statute 6, having erased the name of any Candidate or Candidates for whom he does not vote, and, if he shall have thought fit, having substituted the name of any other Candidate or Candidates contained in the printed list sent in pursuance of Statute 4. One of the Secretaries shall take down the names of the Fellows who vote. No list shall be valid which contains more than fifteen names.

9. The Scrutators, after examining the lists with the Secretaries, shall report to the President the names of the Candidates who shall have been duly elected in compliance with the Charters, and the President shall announce those names from the Chair.

10. Any Candidate whose name shall have been printed in a previous list of Candidates, but who shall not have been elected, shall, if his proposer or seconder so request in writing, before the 31st December immediately preceding an election, be a Candidate at such election; his name shall be placed in alphabetical order with those of the new Candidates to be announced in January following, and his certificate shall be suspended along with those of the new Candidates. Provided always that the same certificate shall not be valid for more than five years from the date thereof. Any separate paper attached to the certificate, if intended to be printed for circulation to the Fellows, shall bear the signatures of six of the signatories to the certificate. Corrections of fact may be made upon the certificate after it has been sent in, if initialled by the proposer and seconder, and marked unmistakably as new matter.

Renewal of Candidature.

11. Any one of His Majesty's subjects who is a Prince of the Blood Royal may be proposed at one of the Ordinary Meetings of the Society by any Fellow, and may be put to the vote for Election on the same day, provided public notice of such proposal shall have been given by the proposer at the preceding Meeting of the Society.

Royal Family.

12. In cases in which the Council is of opinion that, in the interests of the advancement of Natural Knowledge, it is desirable that persons be elected Fellows of the Society otherwise than as provided by Statutes 1 to 11 of this Chapter, they may recommend to the Society for election persons, who, in their opinion, either have rendered conspicuous service to the cause of science, or are such that their election would be of signal benefit to the Society. Provided that not more than two persons shall be so recommended in any one calendar year, and if two persons be elected in any one year, there shall be no election in the following year. The persons

Special Elections.

so to be recommended shall be selected by the Council by ballot. Provided always that no person shall be so recommended unless he obtains two-thirds of the votes of the Members of Council present and voting, the total number of votes registered in his favour being not less than eleven.

At the Ordinary Meeting of the Society next following the Meeting of Council at which such selection is made, the person or persons nominated shall be proposed for election by means of a certificate prepared in accordance with Statute 1, no distinction, however, being made between personal and general knowledge. Such certificate, on being allowed by the Society, shall be suspended in some convenient place in the apartments of the Society until the day on which a ballot is taken upon it. The date for the ballot, which shall not be earlier than the third Ordinary Meeting after that at which the certificate is read, shall be announced at the head of the certificate.

Admission.

13. Every person who is elected a Fellow shall appear for his admission on or before the fourth Ordinary Meeting of the Society after the day of his Election, or within such further time as shall, for some sufficient cause, be granted by the Council; otherwise his election shall be void. Provided that no person shall be admitted a Fellow of the Society on the day of the Anniversary Meeting for electing the Council and Officers.

14. The admission of any Fellow into the Society shall be in manner and form following, he having first made the payments required by the Statutes. Immediately after the reading of the Minutes has been concluded, he shall subscribe the Obligation in the Charter-book, and be introduced to the President, who, taking him by the hand, shall say these words: *I do, by the authority and in the name of the Royal Society of London, for Improving Natural Knowledge, admit you a Fellow thereof.*

Admission, &c., to be recorded.

15. The Election, the payments made previous to admission, and the Admission of every person into the Society, with the time thereof, shall be recorded in the Journal-book.

16. No person shall be deemed a Fellow of the Society until he has made the payments required by the Statutes: nor shall he be entitled to vote at any Election or Meeting of the Society until he shall have been admitted in the manner and form above specified.

Foreign Members.

17. Persons may be elected into the Society, under the title of Foreign Members, who are neither natives nor inhabitants of His Majesty's dominions, and shall be exempted from the operation of

Chapters II and III of these Statutes; they shall be selected from among men of the greatest eminence for their scientific discoveries and attainments.

18. The Council shall from time to time, as they shall see fit, put in nomination persons for Election as Foreign Members, not exceeding, with those already elected, the number of fifty. Procedure for Election of Foreign Members.

19. A book shall be kept in which Fellows of the Society may enter the names of those men of science whom they suggest as Foreign Members; each entry shall be signed by the proposer and be accompanied by a statement of the principal grounds on which the suggestion is made, and shall be valid for three years only.

20. When vacancies in the list of Foreign Members are to be filled up, a list of the persons so entered shall be sent to each Member of the Council, together with notice of the Meeting at which the list will be considered. At the Meeting thus appointed further entries may be made, and the claims of those men of science whose names have been duly entered in the book shall be considered, and a selection of names shall be made, from among which the Council, at a subsequent Meeting to be then appointed, may make nominations to the Society.

21. At the second Meeting the selection of Candidates to be so nominated for Foreign Membership shall be by ballot; when, if two-thirds of the Members of the Council present be in favour of the nomination of any Candidate, his name shall be proposed at the next Ordinary Meeting of the Society, and shall be put to the vote by ballot at the following Ordinary Meeting.

CHAPTER II.

Of the Obligation to be Subscribed.

22. Every person elected a Fellow of the Society shall, before his admission, subscribe the Obligation in the following words:— Obligation to be subscribed by every Fellow.

We who have hereunto subscribed, do hereby promise each for himself, that we will endeavour to promote the good of the Royal Society of London, for Improving Natural Knowledge, and to pursue the ends for which the same was founded; that we will be present at the Meetings of the Society, as often as conveniently we can, especially at the Anniversary Elections, and upon extraordinary occasions; and that we will observe the Statutes and Orders of the said Society. Provided, that whensoever any of us shall signify to the President under his hand, that he desireth to withdraw from the Society, he shall be free from this Obligation for the future.

And if any person elected shall refuse to subscribe the said Obligation, the election of that person shall be void.

CHAPTER III.

Of the Payments to be made by the Fellows to the Society.

Annual Contributions.

23. Every person elected a Fellow of the Society under Statutes 1 to 9, shall, before he is admitted, pay the sum of *ten pounds* for admission money, the sum of *four pounds* for the year of his election, and the same sum annually in advance so long as he shall continue a Fellow of the Society. And if any such person shall fail to pay the said sums, he shall not be admitted, and his Election shall be void: except the said sums be remitted in whole, or in part, by special order of the Council. Provided always that the admission fee of such Fellow shall be paid out of the Fee Reduction Fund, and shall not be demanded of the Fellow; and that, except in the case of Fellows elected before January, 1879, *one pound* of the annual contribution shall be paid out of the Fee Reduction Fund.

A Fellow elected under Statute 12 shall pay no admission money; he shall pay a subscription of four pounds for the year in which he was elected and annually so long as he remains a Fellow of the Society, provided that in the case of Fellows elected after October 1 in any one year and not admitted before the succeeding Annual Meeting, the Council may by special order remit in whole or in part the fee for the year during which the election took place.

Princes of the Blood Royal elected under Statute 11 shall not be required to pay an admission fee or annual subscription.

Life Compositions.

24. All who have or may become Fellows of the Society may at any time compound for their annual payments, by paying at once the sum of *sixty pounds*.

Annual Contributions due March 25th.

25. All Annual Contributions shall be considered to be due on the 25th day of March in each year. Every Fellow of the Society liable to an Annual Payment shall, previously to the 25th day of March in every year, bring or send the same to the Treasurer or the Assistant Secretary. And if any such Fellow, after notice sent by post to his usual address, in May, and again in September, shall fail to pay the same before the first day of October in each year, his name shall be suspended in the public Meeting-room of the Society as being in arrear, and shall continue so suspended until the sum due be

paid. And if any such Fellow shall fail to pay his subscription on or before the first day of November in each year, no satisfactory reason having been assigned to the President and Council for such non-payment, he shall cease to be a Fellow of the Society. Provided, nevertheless, that on a solicitation for readmission being addressed to the President and Council by an individual so circumstanced, within the space of one year following St. Andrew's Day, the case of the individual so soliciting shall be stated by the President from the Chair, at one of the Ordinary Meetings of the Society, and the question of his readmission be put to the vote at the next Ordinary Meeting of the Society. **Consequences of Non-payment.**

CHAPTER IV.

Of the Death or Recess of Fellows.

26. The Death or Recess of any Fellow of the Society shall be forthwith recorded in the Register of Fellows, and the names thus recorded shall be announced from the Chair at the Anniversary Meeting for electing the Council and Officers. **Record of Deaths, &c.**

CHAPTER V.

Of the Causes and Form of Ejection.

27. If any Fellow of the Society shall contemptuously or contumaciously disobey the Statutes or Orders of the Society or Council; or shall, by speaking, writing, or printing, publicly defame the Society; or advisedly, maliciously, or dishonestly do anything to the damage, detriment, or dishonour thereof, he shall be ejected out of the Society. **Grounds for Ejection of Fellows.**

28. Whensoever there shall appear to be cause for the ejection of any Fellow out of the Society, the subject shall be laid before the Council; and if a majority of the Council shall, after due deliberation, determine by ballot to propose to the Society the ejection of the said Fellow, the President shall in that case, at some Ordinary Meeting of the Society, announce from the Chair such determination of the Council; and at the Ordinary Meeting next after that at which the said announcement has been made, the Society shall proceed to determine the question; and on its appearing that two-thirds of the Members present have voted for the ejection of the said Fellow, the President shall proceed to cancel his name in the Register, and at the same time pronounce him ejected in these words:— **Procedure in Ejection of Fellows.**

I do, by the authority and in the name of the Royal Society of London, for Improving Natural Knowledge, declare A. B. to be now ejected, and no longer a Fellow thereof.

And the ejection of every such person shall be then recorded in the Journal-book of the Society; and his name, as ejected, be also read at the next Anniversary Meeting for Elections.

CHAPTER VI.

Of the Election of the Council and Officers.

Notice of Election of Council and Officers.

29. At the two Ordinary Meetings of the Society next preceding the day of the Anniversary Election, the President shall give notice of the said Election; and declare how much it imports the good of the Society, that such persons may be chosen into the Council, as are most likely to attend the Meetings and business of the Council, out of whom there may be made the best choice of a President and other Officers.

Summons to Anniversary Meeting.

30. Every Fellow of the Society resident in the United Kingdom shall have notice of the Anniversary Meeting for electing the Council and Officers for the year ensuing, by particular summons, which summons shall be sent to the address of such Fellow, a week at the least before the day of Meeting, and shall be to this effect:—

> *These are to give notice, that on the day of the Council and Officers of the* Royal Society *are to be elected for the year ensuing; at which Election your presence is expected, at of the clock in the precisely.*

Nomination of the Council and Officers.

31. The Council for the ensuing year, out of which shall be chosen the President, Treasurer, two Secretaries, and Foreign Secretary, shall consist of eleven Members of the existing Council, and of ten Fellows who are not Members of the existing Council.

32. The President and Council shall, previous to the Anniversary Meeting, nominate, by ballot, eleven Members of the existing Council, and also ten Fellows, not Members of the existing Council, whom they recommend to the Society for Election into the Council for the ensuing year. The President and Council shall also, in like manner, nominate by ballot, out of the proposed Council, the persons whom they recommend to the Society for election to the offices of President, Treasurer, two Secretaries, and Foreign Secretary for the ensuing year.

Election.

33. At the Ordinary Meeting of the Society preceding the Anniversary Meeting, the names of such persons so recommended for election as Council and Officers for the ensuing year shall be announced from the Chair.

34. Lists, with the names of the Fellows recommended by the President and Council, and having a blank column opposite for such alterations as any Fellow may wish to make, shall be prepared for the use of the Fellows one week before the day of Election. **Balloting Papers.**

35. Two Scrutators shall be nominated by the President, with the approbation of the Society, to examine the lists. **Scrutators.**

36. Each Fellow voting shall deliver his list to one of the Secretaries or Scrutators; and the name of each Fellow who shall so deliver in his list shall be noted by one of the Secretaries. **Mode of Voting.**

37. The Scrutators, after examining the lists, shall report to the Society the names of those having the majority of votes for composing the Council, and filling the offices of President, Treasurer, the two Secretaries, and Foreign Secretary.

38. For electing any Member of the Council, or any Officer to be elected by the Society, upon such vacancies as shall happen in the intervals of the Anniversary Elections, the summons for such Election, and the proceedings in it, shall be after the same manner as is directed for the Anniversary Election. **Elections to fill Casual Vacancies.**

39. Upon any vacancy of the President's place, occurring in the intervals of the Anniversary Elections, the Treasurer, or, in his absence, one of the Secretaries, shall cause the Council to be summoned for the Election of a new President: and the Council, meeting thereupon in the usual place, or any eleven or more of them, shall proceed to the said Election, and not separate until the major part of them shall have agreed upon a new President.

CHAPTER VII.

Of the President.

40. The business of the President shall be to preside at all the meetings, and regulate all the debates, of the Society, Council, and Committees; to state and put questions both in the affirmative and negative, according to the sense and intention of the meetings; to call for reports and accounts from Committees, and others; to check irregularities, and to keep all persons to order; to summon all Meetings of the Council, and Committee of Papers; and to execute, or see to the execution of, the Statutes of the Society. **Business of the President.**

41. The President shall take precedence of every Fellow of the Society, at their ordinary place of meeting; and also in all other places, where any number of the Fellows meet as a Society, Council, or Committee. **Precedence of the President.**

Vice-Presidents.

42. The President shall annually appoint two or more Members of Council to be Vice-Presidents. In the absence of the President, one of the Vice-Presidents shall act as his deputy, and may do the same acts as the President himself could do if present; provided always that the Council may appoint any Fellow to be Chairman of any Committee other than the Committee of Papers, and in the absence of the President, such Fellow shall act as Chairman of such Committee.

CHAPTER VIII.

Of the Treasurer and his Accounts.

Duties of the Treasurer.

43. The Treasurer, or some person appointed by him, with the approval of the Council, shall receive for the use of the Society all sums of money due or payable to the Society; and shall pay and disburse all sums due from or payable by the Society; and shall keep particular Accounts of all such receipts and payments.

Power to make Payments.

44. Every sum of money, payable on account of the Society, exceeding Ten Pounds, shall be paid only by order of the Council; but payments for rates or taxes, to any amount, may be made by the Treasurer, without any specific order of the Council for that purpose.

Investment of Moneys.

45. All sums of money, which there shall not be present occasion for expending or otherwise disposing of to the use of the Society, shall be laid out in such Government or other securities as shall be approved of and directed by the Council.

Audit of Accounts.

46. The Accounts of the Treasurer shall be audited annually, a short time preceding the Anniversary Elections, by a Committee consisting of three Members of the Council, of whom the President or one of the Secretaries to be one; and of three Fellows of the Society not Members of the Council, who are to be nominated by the President, with the consent of the major part of the Fellows present, given by ballot at one of the three next preceding weekly meetings; any one or more of the said three Members of the Council, together with any one or more of the said three Fellows, shall be a Quorum of the said Committee: the Members of the said Committee who are of the Council shall make their Report to the Council held next after such audit, on or before the Anniversary Election; and the Members of the said Committee who are not of the Council shall make their Report to the Society, upon the Meeting next before the Anniversary Election, or on the day of the said Election.

47. As soon after the Audit as may be, and before the Anniversary Meeting, the Treasurer shall cause an abstract of the Society's Accounts of the preceding year to be printed for the use of the Fellows. Accounts to be printed.

48. The Treasurer shall have the charge of the Title Deeds of the Society's Estates, the Policies of Insurance, and Securities, and shall have the general care of the Society's property. Charge of Title Deeds.

CHAPTER IX.

Of the Two Secretaries.

49. The two Secretaries shall be responsible to the Council for the general conduct of the Society's correspondence, publications, and all other business, excepting that which relates to finance. Duties of the Secretaries.

50. The Secretaries, or one of them, shall attend all meetings of the Society, Council, and Committee of Papers; where, when the President has taken the Chair, one of the Secretaries shall read the minutes, orders, and entries of the preceding meeting; and shall be responsible for the accuracy of the minutes of the business and orders of the present meeting, to be entered by the Assistant Secretary in the respective books to which they relate.

CHAPTER X.

Of the Foreign Secretary.

51. The duty of the Secretary for Foreign Correspondence shall be to conduct correspondence with foreign parts, relating to the business of the Society, to return thanks for presents from foreigners made to the Society, and to forward to persons elected Foreign Members the Diplomas certifying their election into the Society. Duties of the Foreign Secretary.

CHAPTER XI.

Of the Assistant Secretary.

52. The appointment of a person to the office of Assistant Secretary shall be by the Council, to whom the Officer so appointed shall give security, at the discretion of the Council; and he shall hold office during the pleasure of the Council. Appointment of the Assistant Secretary.

Assistant Secretary not a Fellow.

53. The person who shall be chosen to the office of Assistant Secretary, shall either not be a Fellow of the Society or, if a Fellow, shall cease to be so upon his appointment to, and acceptance of that office.

Payment of the Assistant Secretary.

54. The Assistant Secretary shall be paid for his services according to the determination of the Council. He shall be subject to such Standing Orders as shall be made, and shall follow such directions as may be given by the President and Council. He shall give all his time to the work of the Society, and shall attend in the Office of the Society as prescribed by such Standing Orders.

55. The Assistant Secretary shall act under the general direction of the Secretaries, and in their absence shall represent them.

He shall also act on the directions which may be given him from time to time by the Treasurer in that part of his duties which relates to the finances of the Society.

Duties of the Assistant Secretary.

56. Subject to such Standing Orders and directions as aforesaid, the Assistant Secretary shall be charged with (1) the general administrative business of the Society: (2) the control of the Offices and library and of all persons therein employed, and the arrangement of the office work: (3) the conduct of the general correspondence.

It shall be his duty, except on such special occasions as may be specified by Standing Orders, or as may be determined by the President, to attend all meetings of the Council and Committees and take the minutes thereof, which minutes he shall submit to the Secretaries for their revision or approval.

57. He shall, subject as aforesaid, have the charge and custody of the Charter-book, Statute-book, Journal-books of the Society and Council, Register-books, Manuscripts, and Archives belonging to the Society; all which shall, except in such cases as the Council shall otherwise order, be kept in the House of the Society, that they may be in readiness to be produced at any meetings of the Society or Council, as the case may require, or as shall be ordered by the Society, Council or President.

58. He shall also, subject as aforesaid, have the charge and custody of the papers and writings relating to the business of the Society, and shall not suffer them to be removed from the House of the Society without the sanction of the Officers.

CHAPTER XII.

Of the Ordinary Meetings of the Society.

59. The Session of the Society shall commence on the first Thursday in November, and end on the last Thursday in June. **Duration of Session.**

60. The Ordinary Meetings of the Society shall be on Thursdays weekly (excepting Christmas, Passion, and Easter weeks, and such other weeks at Christmas and Easter, in each year, as the Council may in the preceding year determine) and shall begin at half-past Four o'clock in the Afternoon precisely. **Days of Meeting.**

61. No stranger shall be permitted to be present during the Meeting, unless by invitation of the President, or by his leave or order upon the recommendation of some Fellow. **Admission of Strangers.**

62. The business of the Society in their Ordinary Meetings shall be to order, take account, consider, and discourse of philosophical experiments and observations; to read, hear, and discourse upon letters, reports, and other papers containing philosophical matters; as also to view, and discourse upon, rarities of nature and art: and thereupon to consider, what may be deduced from them, or any of them; and how far they, or any of them, may be improved for use or discovery.* **Business of Ordinary Meetings.**

63. At the meetings of the Society, Lists of the Presents made from time to time to the Society shall be laid on the Table, by one of the Secretaries, for the inspection of the Fellows; and the thanks of the Society to the Donors shall be proposed from the Chair previously to the reading of the first Paper. One of the Secretaries shall give notice of any Candidate who stands proposed for election into the Society at that Meeting; and the Secretaries shall read Letters and Papers presented to the Society, in such manner as the President shall direct. **Procedure at Meetings.**

64. No letter, report, or other paper shall be read at any Ordinary Meeting unless it be communicated by a Fellow or Foreign Member; and it shall be the duty of each Fellow or Foreign Member to satisfy himself that any letter, report, or other paper which he may communicate, is suitable to be read before the Society. **Communication of Papers.**

65. The conduct of the Ordinary Meetings shall be in accordance with the Standing Orders determined from time to time by the **Conduct of Meetings.**

* This is the wording of the Statute as given in the Statutes of 1663.

President and Council, provided always that at the Ordinary Meetings nothing relating to Statutes or management of the Society shall be brought forward or discussed.

CHAPTER XIII.

Of Special General Meetings of the Society.

Anniversary Meeting.

66. The Anniversary Meeting for the election of the Council and Officers, and the Annual Meeting for the election of Fellows, shall take place at an hour to be determined by the Council.

Special General Meetings.

67. The President or Council may at any time call a Special General Meeting of the Society when it may appear to them to be necessary.

68. Any six Fellows may, by notice in writing, signed by them, and delivered to one of the Secretaries at an Ordinary Meeting of the Society, require a Special General Meeting of the Society to be convened, for the purpose of considering and determining on the matters specified in such requisition, and the President shall, within one week after such requisition shall have been so delivered, appoint a day for a Special General Meeting accordingly.

69. One week's notice of any Special General Meeting shall be given to each Fellow resident in the United Kingdom, and such notice shall state the object of such Meeting.

70. At such Meeting no business shall be brought forward except what shall have been so notified.

71. A Special General Meeting of the Society shall be held before the Annual Meeting on the Thursday preceding the Annual Meeting. At such Special Meeting the Report of the Council, having been previously circulated, shall be presented, and Fellows shall be invited to comment thereon. Matters not arising out of the Report shall not be discussed except by permission of the President. The Report shall be finally presented to the Society at the Anniversary Meeting, but no change shall be introduced after its first presentation other than such as the Council may think fit to adopt on consideration of the comments made at the Special General Meeting.

CHAPTER XIV.

Of the Publication of Papers.

Committee of Papers.

72. The Members of the Council for the time being shall constitute and be a standing Committee, to be called the Committee of Papers, to whom the consideration of the acceptance, reading, and publication

of all papers communicated to the Society shall be referred, and who shall execute their powers in accordance with Standing Orders determined from time to time by the President and Council. The Committee of Papers shall meet at such times as shall be appointed by the President; due and sufficient notice of such meeting having been previously sent to every Member of the Committee.

73. At a meeting of the said Committee no less number than seven of the Members (of which number the President, or, in his absence, a Vice-President, shall always be one) shall be a quorum. **Quorum.**

74. The decisions of the Committee of Papers shall be determined by the majority of votes of those present and voting, and the voting shall be open, unless the President shall direct that the voting shall be by ballot. In case of an equality of votes, the President shall have a second or casting vote. **Voting.**

The decisions of the Committee shall be duly entered in the Minute-book of the Committee.

75. The publication of papers communicated to the Society, and of such other matters as the President and Council may judge fit to publish, shall take place under Standing Orders determined from time to time by the President and Council, but always in such a way that a proper portion of them shall from time to time be printed and published under the title of the 'Philosophical Transactions of the Royal Society of London,' and another proper portion under the title of the 'Proceedings of the Royal Society of London,' provided always that the President and Council shall have power to publish either papers or other matter in such form and under such conditions as they may from time to time determine. **Form of Publication.**

76. The original copy of every Paper received at the Society shall be considered the property of the Society, if there be no previous engagement with its author to the contrary. **Papers communicated.**

77. The *Philosophical Transactions* and the *Proceedings* shall be printed at the sole charge, and for the use and benefit, of the Society, and of the Fellows thereof; to the intent that each of the present Fellows, who actually contributes and pays towards the support of the Society, or who has compounded for such contribution, according to the rules and orders established in relation thereto, or who has for other particular reasons been exonerated and discharged from such contribution by order of the Council, may receive *gratis*, under proper limitations, one copy of such of the *Philosophical Transactions*, and of the *Proceedings* as shall be printed as aforesaid; and that all persons who shall hereafter be admitted Fellows shall, under the **Fellows entitled to Publications *Gratis*.**

same conditions, receive, and be entitled to, the like benefit and advantage.

Delivery of *Transactions.*

78. The Assistant Secretary shall deliver *gratis* one of the said copies of the *Transactions* to every Fellow of the Society (except as hereinafter excepted) who shall demand the same, either in person, or by letter.

Provided always, that no Fellow whatsoever of the Society shall be entitled to demand or receive any such copy of the *Transactions*, whose election and payment of Admission fees and regular Contributions shall not have preceded the date of the time appointed for the delivery of the said *Transactions;* neither shall the Executor of any deceased Fellow receive a copy of the *Transactions* published after the death of such Fellow.

Limitation of Time for claiming *Transactions.*

Provided also, that no Fellow of the Society shall receive, or be entitled to receive, *gratis*, any copy or copies of the *Transactions*, so printed as aforesaid, after five years shall have elapsed from the time of the Assistant Secretary's having begun to deliver out such copies respectively; but his neglecting to demand them for so long a time shall be deemed a forfeiture and dereliction of his right thereto: unless the Council for the time being, upon being made acquainted with the reason of such delay, and having regard to the circumstances of the application, and the amount of stock in hand, shall *order* such copies as they may think fit to be so delivered.

Delivery of *Proceedings.*

79. The Assistant Secretary shall further cause to be distributed *gratis* to all the Fellows of the Society, by post or otherwise, copies of the *Proceedings* as soon as may be convenient after their appearance.

CHAPTER XV.

Of the Books and Papers of the Society.

Charter Book.

80. There shall be had and kept a Book, called the *Charter-book*, wherein shall be fairly written the copy of the Charters, all the Royal Grants on the behalf of the Society, and the Obligation to be subscribed by the Fellows of the Society in their own hand-writing.

Statute Book.

Register of Fellows.

81. There shall be kept a Book, called the *Statute-book*, wherein shall be fairly written, or printed, all the Laws, Statutes, and Constitutions made, or to be made, concerning the government and regulating of the Society or Council; and also a Register of the Fellows of the Society, with the times of their Election and Admission.

82. There shall be kept *Journal-books* of the Society, and also of the Council, wherein shall be entered all the minutes, orders, and business of the Society and Council at their respective meetings; to which Journal-books any Fellow may have access at such times as the Library is open. Journal Books.

83. A Book shall be kept, in which the title of each communication received, the date of its reception at the apartments of the Society, and the name of the Fellow or Foreign Member who communicates it, shall be duly entered in the order of its reception. Register of Papers.

CHAPTER XVI.

Of the Archives of the Society.

84. The scientific and historical manuscript collections of the Society, including the collections of engravings, and also confidential reports relating to papers submitted to the Society, shall be preserved in the Archives, and catalogued so as to be available to duly authorized persons for consultation and study. The manuscripts of all papers read before the Society, of which the publication has been deferred, shall be preserved in the Archives. Manuscripts.

85. A catalogue of the manuscripts in the Archives shall be available for reference at the rooms of the Society; and permission may be granted by the Council or Officers of the Society to Fellows or to any person duly introduced by a Fellow, to consult and examine specified manuscripts not of a confidential character. Catalogue.

86. No manuscript deposited in the Archives shall be removed from the rooms of the Society without permission of the Council, on security given for its safe custody and return. Manuscripts not to be removed.

CHAPTER XVII.

Of the Library.

87. The Library shall be open to the Fellows on such days, and at such hours, and subject to such regulations, as the Council may from time to time determine by Standing Orders. Library Regulations.

CHAPTER XVIII.

Of the Common Seal and Deeds.

Care of the Seal.

88. The Common Seal of the Society shall be kept in a box, the key of which shall be kept in a sealed packet. When the Common Seal has to be used, this packet shall be opened by the President in Council; and at the Council meeting at which it is so opened, the Common Seal having been replaced in the box, and the box locked, the key shall again be enclosed in a packet, which shall be sealed by the President with his private seal. The box and sealed packet shall be kept at the Society's Apartments in an iron safe.

Sealing Deeds.

89. Every Deed or writing, to which the Common Seal is to be affixed, shall be passed and sealed in Council.

CHAPTER XIX.

Of the Restraint of Dividends to Fellows.

90. The Society shall not, and by its laws may not, make any Dividend, Gift, Division, or Bonus in Money unto or between any of its Members.

CHAPTER XX.

Of the Standing Orders of the Society.

Power to make and alter Standing Orders.

91. The Council may from time to time make Standing Orders for the regulation of the affairs of the Society, provided that such Standing Orders be not contrary to anything contained in the Charters or Statutes of the Society. Such Standing Orders may be made, amended or repealed at any meeting of the Council, provided that notice of any proposed new Standing Order, amendment or repeal has been given at the preceding Council. Such notice shall specify the proposed new Standing Order, amendment or repeal.

CHAPTER XXI.

Of the Making, Amending, and Repealing of Laws.

Procedure in making or altering Statutes.

92. For the making of any new Statute of the Society or for the amendment or repeal of any Statute or any part thereof, notice of the proposed new Statute, amendment or repeal shall be given at a meeting of the Council, and such notice shall specify the proposed

new Statute, amendment or repeal. The Council shall thereupon appoint a day for the consideration of the proposal. On such day the question to be resolved shall be to this effect, viz., "Whether the proposed new Statute, amendment or repeal shall be proposed at another meeting?" And if this be determined in the affirmative the Council shall appoint a day for such proposal. On such day the question shall be to this effect, viz., "Whether the proposed new Statute, amendment or repeal shall be agreed upon?" And in case the said new Statute, amendment or repeal be agreed unto, the same shall be recorded in the Statute Book of the Society, and when so recorded shall pass for a law.

March 18, 1915.

NOTE.

By the Charter, the quorum required is as follows:—

Ordinary Meeting of the Society	9
Anniversary Meeting for Election of Council	31
Meeting to elect to casual vacancies on Council	21
Meeting for Election of Fellows	21
Meeting of Council	9

STANDING ORDERS OF COUNCIL RELATING TO MEETINGS OF THE SOCIETY, SECTIONAL COMMITTEES, AND PUBLICATIONS.

(*As amended May* 21, 1914.)

NOTE.

By Statutes, Cap. XIV, Stat. 72, the consideration of the acceptance, reading, and publication of all papers communicated to the Society is referred to the Council sitting as Committee of Papers; and in the following Standing Orders the word "Council," when used in connexion with the acceptance, reading, or publication of papers, is to be understood to mean the Council sitting as Committee of Papers.

Relating to the Conduct of Ordinary Meetings.

1. At each Ordinary Meeting, any formal business of the Society which may be necessary, such as the reading of certificates, balloting for candidates under Statutes, Cap. I, announcements, returning thanks for presents, &c., shall, unless the President direct otherwise, be the first business of the meeting.

2. At each Ordinary Meeting, not being "a Meeting for Discussion," as hereinafter provided, or for the Bakerian or the Croonian Lecture, the President shall determine what papers are to be read, and the order in which they shall be taken. He may also, whenever he sees fit, direct the author of a paper or one of the Secretaries to read an abstract of the paper or the paper itself, if it be sufficiently brief, or may invite the author to make an oral statement of the nature of its contents, and may also invite remarks upon the paper. When an oral statement is desired, the author shall, so far as possible, be previously informed of the fact. A paper shall be considered to have been "read" if one of the Secretaries has read its title only.

3. At any Ordinary Meeting, not being a "Meeting for Discussion," any Fellow of the Society may, with the approval of the President, and at such period of the Meeting as the President may determine, make a communication not of the nature of a "paper," or exhibit objects having relation to the advancement of Natural Knowledge.

4. The President shall further have power at any Ordinary Meeting, and at any period of that Meeting which he may think proper, to make such announcements or statements, as he may think desirable, relating to the advancement of Natural Knowledge.

5. In each year certain Ordinary Meetings, not more than four in number (exclusive of the Meetings set aside for the Bakerian and Croonian Lectures respectively), shall be devoted each to the hearing and consideration of some one important communication, or to the discussion of some important topic; these Meetings shall be termed "Meetings for Discussion."

6. The Council shall from time to time give due notice of the dates at which Meetings for Discussion will be held.

7. The Council, of its own motion, or upon the recommendation of a Sectional Committee, may select some communication made to the Society in the ordinary way, as the subject for such a Meeting for Discussion, or it may select for that purpose some question, the discussion of which would, in their judgment, be likely to advance Natural Knowledge. In the latter case, the Council shall appoint some person to open the discussion by means of a communication made by him for that purpose.

8. When a Meeting for Discussion has been arranged, the Council, or the Officers, shall direct printed copies of the communication which has been approved of for the said Meeting (or of an adequate abstract of it), to be sent not later than one week before the date of the Meeting, to each Fellow, or to certain Fellows of the Society, and to such other persons as the President may direct. And the Council shall take such other steps as may seem to it desirable to render the discussion useful towards the advancement of Natural Knowledge.

9. At each Meeting for Discussion, the conduct of the discussion shall be under the direction of the President, who shall arrange for the Fellows present and desiring to speak, and who shall have the power to invite, if he think fit, persons present, not Fellows of the Society, to take part in the discussion. Any Fellow shall be at liberty to send to the Secretaries, previous to the Meeting, written remarks on the communication which is the subject of the meeting, and the President shall, if he see fit, direct one or other of the Secretaries to read these remarks at the meeting.

Relating to Sectional Committees.

10. The Council shall appoint, from among the Fellows of the Society, Committees representing the several branches of Natural Knowledge, and called "Sectional Committees." The Members of

each Committee shall be chosen with a view to secure, so far as is possible, a representation of the several sub-divisions of each branch of Natural Knowledge, and to obtain the assistance of Fellows who, from their connexion with other societies, and otherwise, are specially qualified to advise the Council in respect to particular parts of Natural Knowledge.

11. Each Sectional Committee shall advise the Council (whether sitting as the Committee of Papers or otherwise) or any of the Officers upon matters referred to it by the Council or by any of the Officers, and otherwise shall make to the Council such suggestions as it may think desirable touching the branch or branches of Natural Knowledge which it represents; it being understood that no Sectional Committee shall offer advice to the Council as to the selection of candidates for admission into the Society as Fellows or Foreign Members, or as to the awards of Medals, unless the Council shall have asked for such advice.

12. The Council shall each year appoint a Member of each Committee to serve as Chairman of that Committee, and to be the channel of communication between the Committee and the Council or Officers. He is authorized to appoint one of the Committee his Deputy when necessary.

13. The Sectional Committees shall be six in number, viz.:—

(1) A "Mathematics" Committee for Mathematics, Mathematical Physics, Crystallography, and Mathematical Astronomy.

(2) A "Physics and Chemistry" Committee for Experimental Physics, Observational Astronomy, Meteorology, Chemistry, and Metallurgy.

(3) A "Geology" Committee for Geology, Palæontology, Mineralogy, and Geography.

(4) A "Botany" Committee for Botany.

(5) A "Zoology" Committee for Zoology and Comparative Anatomy.

(6) A "Physiology" Committee for Animal Physiology and Medical Subjects.

14. The "Mathematics" Committee shall consist of six Members, of whom two shall retire each year; three Members shall form a quorum.

The "Physics and Chemistry" Committee shall consist of twelve Members, of whom four shall retire each year; five Members shall form a quorum.

The "Geology" Committee shall consist of nine Members, of whom three shall retire each year; four Members shall form a quorum.

The "Botany" Committee shall consist of nine Members, of whom three shall retire each year; four members shall form a quorum.

The "Zoology" Committee shall consist of nine Members, of whom three shall retire each year; four members shall form a quorum.

The "Physiology" Committee shall consist of twelve Members, of whom four shall retire each year; five members shall form a quorum.

It shall be in the power of the Council to add to the number of any Committee, if at any time it may seem to be desirable to do so.

15. Any Member of Council who desires to attend the meetings of any Sectional Committee, of which he is not at the time being a Member, shall have power to do so as *amicus curiæ* under the following conditions. Upon his expressing in writing to the Assistant Secretary his wish so to attend, the summons for each meeting of the Committee shall be sent to him as to an ordinary Member of the Committee during his tenure of office as Member of Council, or during such shorter time as he may name; but the Chairman of the Committee shall not be expected to correspond with him as with an ordinary Member of Committee. He may with the consent of the Chairman speak during the deliberations of the Committee, but shall give no vote.

16. The retirement of Members shall be determined by seniority of appointment to the Committee.

17. The retiring Members of the Committee shall each year vacate office on the 31st of December, and shall not be eligible for election for the ensuing year.

18. Should, by reason of death or otherwise, a vacancy occur at any intermediate time, the Council shall appoint a person to fill the vacancy, and the retirement of the person so appointed shall be according to the rules which would have applied to the Member whose place he fills, provided that, if at the date of retirement the said person has not served more than one year, he shall be eligible for immediate re-appointment.

19. The appointment of the Fellows to serve as new Members of Committee shall be made by the Council in December, and the Members so appointed shall enter office upon the 1st of January ensuing.

20. Each Committee shall hold a Meeting in January every year, at which they shall make arrangements for their Meetings during the ensuing year, and determine the procedure to be adopted in exercising the control of the Committee over the publication of papers.

21. The decisions arrived at by a Meeting of the Committee at which the Members present do not form a quorum shall be valid, if

subsequently agreed to in writing by not less than two-thirds of the whole Committee.

22. At each Meeting of the Committee the Chairman shall submit (*a*) a list of papers which have been provisionally approved for publication under Standing Order 43; (*b*) a list of papers which have been accepted for reading under Standing Order 39, but with regard to which no steps towards publication have been taken; (*c*) a list of papers communicated, but not so far accepted for reading. The Committee shall take such steps as they think fit with regard to the papers submitted to them. In the case of papers approved by the Committee, the Secretaries shall proceed to publication immediately.

23. Voting shall be open unless any member of the Committee demand a ballot.

24. Papers which have been approved by a majority of the Members of a Committee by correspondence shall be considered approved by the Committee, and be treated as such in the statement laid by the Secretaries before the Council.

25. The Minutes of the Committee shall be duly recorded in a book kept for that purpose, and preserved in the apartments of the Society, together with such correspondence and documents relating to the business of the Committee as the Committee may think it desirable to preserve.

26. When a Committee is of opinion that a paper referred to it might profitably serve as the basis of a discussion at a meeting of the Society, it shall forthwith report to that effect to the Council. If the matter seem urgent, the President and Officers shall have power, without waiting for a Meeting of the Council, to take immediate steps towards carrying out the recommendations of the Committee.

27. Should at any time a Committee be of opinion that it would be desirable to encourage a discussion at a meeting of the Society upon some subject, concerning which no paper suitable to serve as a basis for discussion is under its consideration, the Committee may invite some person to prepare a paper suitable for that purpose, and may recommend to the Council to treat such a paper as the basis of a discussion.

28. The Chairmen of Sectional Committees, together with the Officers of the Society, shall be a Standing Committee with powers to make such recommendations to the Council as they think fit concerning questions affecting the Sectional Committees jointly, or on any matters affecting the Society and not specifically excluded from their purview by resolution of Council. They shall meet at least once a year, and elect a Chairman who shall act as convener.

As to 'Proceedings,' 'Philosophical Transactions,' and 'Year-book.'

29. The 'Proceedings' shall be the medium of publication for papers of approved merit not more than twenty-four pages in length, and not containing numerous or elaborate illustrations, provided that the Council may in exceptional circumstances, on a report from a Sectional Committee, order papers of greater length to be printed in the 'Proceedings.' The 'Proceedings' shall also contain short abstracts of papers to be published in the 'Philosophical Transactions' (*see* Standing Order 47), and such other matters as the Council may judge desirable to insert.

30. The 'Philosophical Transactions' shall be the medium of publication (*a*) for papers of approved merit which contain numerous or elaborate illustrations, or which cannot without detriment to their scientific value be condensed into the space reserved for papers in the 'Proceedings'; (*b*) for such reports or investigations carried out by a Committee of the Society as the Council may order to be printed in the 'Transactions.'

31. The 'Proceedings' shall be issued in two series—(A) containing those papers which are of a mathematical or physical character, and (B) containing those of a biological character.

32. Subject to the powers given to the Secretaries and Chairmen of Sectional Committees in Standing Orders 39 and 42, it shall rest with the Council, after a report of the Sectional Committee, whether 'Proceedings' or 'Philosophical Transactions' is the more suitable medium for publication. The Communicator of a paper shall, however, be at liberty to indicate his preference for one or other of the two forms of publication.

33. The 'Proceedings' shall be published in numbers issued at as short intervals as may be found suitable.

34. Each paper ordered by the Council for publication in the 'Philosophical Transactions' shall be published separately in paper covers, the date at which it is issued being marked on the cover; and shall be sold separately. The papers shall be issued in two series, viz., series A containing papers of a physical or mathematical character, and series B containing those of a biological character. At suitable intervals the papers in each series shall be collected into a volume.

35. A Year-book of the Society shall be published annually, containing :—

A list of Fellows of the Society and their addresses;

A list of the Members of Standing Committees, and of the Government Grant Boards;

The Statutes of the Society;
The Standing Orders of Council;
Additions to the Library in each year;
The awards of Medals;
The Balance Sheet for the previous year;
The minutes of Meetings of the Society;
And such other matters as the Council may order to be inserted.

Relating to acceptance, reading and publication of papers.

36. No paper is received by the Society unless it be communicated by a Fellow or Foreign Member, who shall satisfy himself that the paper is a fit and proper one to be communicated to the Society, and has not been previously published elsewhere. (Statute 64, Chapter XII.)

37. Each paper must be accompanied by a summary not exceeding 300 words in length, showing the general scope of the communication, and indicating points which, in the opinion of the author, are of special importance. In the case of papers accepted for reading, this summary is printed for the use of the Fellows at the Meeting at which it is read, and afterwards in the report of the Meeting.

38. Upon a communicated paper reaching the apartments of the Society, the Assistant Secretary shall mark on it the date of the reception and record it in the Register of Papers (Statute IV, Chapter 15). The Assistant Secretary shall report the reception of the paper to the one or the other of the Secretaries, according to the nature of the communication, and to the Chairman of the appropriate Sectional Committee.

39. The Secretary to whom a paper is reported (*see* Standing Order 38) shall, if he thinks fit, of himself or after consultation with the Chairman of the appropriate Sectional Committee, direct the paper to be marked as "accepted for reading." If the Secretary, after consultation with the Chairman, is of opinion that further advice is desirable, the question of acceptance for reading shall be referred to the appropriate Sectional Committee. In this case the Chairman of the Committee shall, at a meeting of his Committee, or by correspondence with its Members, obtain the view of the Committee thereon, and report the same to the Secretary, who shall, if the decision be favourable, act on the advice so given; otherwise he shall report the view of the Committee to the Council.

40. In the case of a paper not being accepted for reading, the Fellow communicating the paper shall be informed thereof as soon as the Council have confirmed the decision of the Sectional Committee. A paper which has not been read may be "withdrawn," provided

that the paper, or a certified copy thereof, remains in the possession of the Society. The Assistant Secretary shall mark clearly on such paper or copy that it has been withdrawn.

41. If after a paper has been read, but before it has been passed for "Press," the author or communicator shall indicate in writing his wish that the paper be not published, his request shall be laid before the Sectional Committee, and if assented to, publication of the paper shall not be proceeded with; but the paper, or a certified copy thereof, shall remain in the possession of the Society.

42. When a paper has been accepted for reading, the author or communicator of the paper shall be informed as soon as the meeting at which it is to be read has been determined upon.

43. In the case of papers accepted for reading, and not exceeding in length twenty-four pages of either the 'Proceedings,' or the 'Philosophical Transactions,' the Secretary, after consultation with the Chairman of the appropriate Sectional Committee, may, if he thinks fit, proceed to publication, but the Sectional Committees shall have power to make such regulations as they may desire, to ensure that no paper shall be finally passed for "Press" without an opportunity having been given to the Members of the Committee to express an opinion on the paper, at a meeting or by correspondence. Before exercising their power to proceed to publication, the Secretary and Chairman may obtain the opinion of one or more persons with regard to the merits of the paper.

44. Papers with regard to which the Secretaries have not exercised the power given to them in Standing Order 43, and papers exceeding twenty-four pages in length, shall be referred to the appropriate Sectional Committee through its Chairman, the Secretary and Chairman having in the meantime, if they think fit, obtained the opinion thereon in writing of one or more persons who are knowing and well skilled in the particular branch of Natural Knowledge to which the paper relates.

45. In the case of the Chairman of a Sectional Committee being the author or communicator of a paper referred to that Committee, the Secretaries shall have power in consultation with some Member or Members of the Committee other than the Chairman, to take the same action as under the foregoing Standing Orders they are empowered to take in consultation with the Chairman.

46. The Sectional Committee shall, with regard to any paper brought to its notice, express its opinion whether and with what modifications, if any, the paper shall be published, and whether the paper shall be published in the 'Proceedings' or in the 'Philosophical Transactions,' provided (*a*) that no paper exceeding twenty-four pages in length shall be recommended for publication in the 'Philosophical

Transactions' without its having been submitted to two Referees, and (*b*) that no paper exceeding twenty-four pages in length shall be recommended for publication in the 'Proceedings,' except for special reasons, which shall be specified in the Report of the Sectional Committee to the Council.

47. In the case of papers accepted for publication in the 'Philosophical Transactions,' the author shall be at liberty to furnish a short abstract for publication in the 'Proceedings.'

48. Should the Secretaries be of opinion that a paper pertains by its subject to more than one Sectional Committee, they shall take steps to obtain the judgment of the several Committees concerned. In the case of any difficulty as to the reference of a paper to its appropriate Sectional Committee or Committees, the Secretary concerned shall bring the matter before the Council.

49. In the case of a paper being submitted to a referee as regards acceptance for reading or publication, his opinion in writing shall be asked on the following points :—

i. Whether or not the paper should be read before the Society.
ii. Whether the paper should or should not be published by the Society.
iii. Whether, in the former case, it should be published in the 'Philosophical Transactions' or the 'Proceedings.'
iv. Whether it should be published in full or only in part, or whether any material modifications are necessary.
v. Which illustrations, if any, accompanying the paper shall be produced.

50. The opinion of the referee shall be regarded as confidential by the Society.

51. If the opinions of two referees disagree, and it is found desirable to submit the paper to a third referee, the Secretaries shall have power to communicate to him the grounds on which the original referees have expressed their respective opinions.

52. The attention of the referees shall be drawn to the fact that the body or bodies authorizing publication are not responsible for the opinions expressed by the authors. The referees should be guided in the expression of their opinion by the consideration that the papers published by the Society should either (*a*) contain some material contribution to scientific knowledge, or (*b*) raise points of especial interest the discussion of which is likely to lead to an advance of knowledge. If these conditions are fulfilled, the paper may be recommended for publication, though the conclusions drawn by the author may not be fully established An adverse report, on the other hand, may be given on the ground that the paper is one elaborating points of detail

or dealing with matters which may more suitably be published through other channels.

53. Each paper published by the Society shall bear the date at which it was received by the Society. No addition or change other than mere verbal corrections shall be made subsequently either in the manuscript or in correcting the proofs for press without such changes being clearly marked and dated.

54. If after the decision of a Sectional Committee to recommend a paper for publication, the Secretaries shall, from any discussion that may take place at the meeting at which the paper is read, or otherwise have good reason to think that the Committee should have the opportunity of reconsidering their decision, they shall, after consultation with the Chairman of the Committee, have power to postpone the publication of the paper. In that event the Chairman shall either, at a meeting or by correspondence with the Members, submit the question of publication again to the Committee.

55. The Council shall, at some convenient time during each of its meetings, sit as a Committee of Papers (Statute 72, Chapter XIV), and the summons to the meeting of the Council shall be considered sufficient notice for such meeting of the Committee of Papers.

56. At each meeting of the Committee of Papers a full statement shall be laid on the table giving a list of all papers communicated to the Society and not yet published or withdrawn, with the recommendations made as to publication. The Committee of Papers shall confirm, or otherwise, the recommendations of the Sectional Committees.

57. The attention of the Committee shall be especially drawn to any papers

(1) of which the cost of publication exceeds a sum to be fixed from time to time by the Council.
(2) which are not recommended for publication and have not been withdrawn;
(3) which, exceeding twenty-four pages in length are recommended to be published in the 'Proceedings.'

58. The Secretaries shall report, and the Committee shall take action in the case of any paper (1) which has been referred to two Sectional Committees whose recommendations as to publication are not in agreement, and (2) where the author or communicator does not accept the decision of the Sectional Committee with respect to the medium of publication (Standing Order 32).

59. The Committee of Papers, at each meeting, shall formally report its decisions to the Council, who shall enter such reports in its Minutes in such form as it may from time to time determine.

PROCEDURE IN THE NOMINATION OF THE COUNCIL.*

1. The subject of the new Council shall be taken into consideration at the first Meeting of Council to be held in October; and with the summons for that Meeting there shall be transmitted a list of the Members of the existing Council, with the number of their attendances at Meetings up to that date; also a List of the Fellows of the Society, with an indication of those who have at any time served on the Council, and the dates of their service.

2. At the second Meeting of Council held in October the names of those Members of the existing Council who retire at the ensuing Anniversary shall be determined.

The retirement of members of Council will be determined upon in the following order; with the proviso that every member of Council retains the right to exercise his vote according to his discretion, subject only to the terms of the Charter, which ordains that ten members of Council, and no more, shall be changed at each Anniversary:—

(i) Members who have expressed their wish to retire; and those who have served during the preceding two years, excepting the Officers for the time being.

(ii) Former Officers of the Society who have continued to serve for one year after vacating office.

(iii) Members who have been absent from more than two meetings since the last Anniversary, in the order of least attendance.

(iv) Members who have served on the Council for the first time during the preceding session; provided that if the number of members who would vacate under the preceding categories be greater than ten, the necessary adjustment shall be made by re-nominating a sufficient number of members who are serving for the first time, in the order of their seniority as Fellows.

(v) If the number of members who vacate under the above categories should be less than ten, the required vacancies shall be made up by the retirement of members who have served on a previous Council, excepting Officers for the time being; such retirements to take place in the order of least seniority as Fellows.

(Seniority shall be counted by the date of election into the Society, and, in the case of Fellows elected during the same year, by the order of their admission into the Society.)

* From Minutes of Council, June 20, 1872, and May 21, 1914.

(vi) In cases where any member of the Council considers that the re-nomination of a member is—for special reasons—desirable, although such member would not be re-nominated according to the above rules, a ballot shall be taken on a motion to re-nominate such member, notice having been given at a previous meeting, and the question of re-nomination shall be determined by the vote of the majority of the members of the Council present and voting.

Thereafter each Member present shall hand to one of the Secretaries a List of not exceeding ten Fellows whom he proposes for the new Council, of whom five shall not have already served on the Council. Members not able to be present may send in similar lists previous to the Meeting. The several lists of names so proposed shall then be read out by the Secretaries.

3. Before the next following Meeting, the President and other Officers shall prepare a list of twenty-one names for consideration by the Council, which list shall include ten names selected from those proposed at the previous Meeting, or other names, if required to make up that number. The list so prepared, together with a statement of the names proposed, and the number of votes given for each, shall be sent out confidentially with the summons for the ensuing Meeting, at which Meeting the names to be finally recommended shall be balloted for. In taking the ballot, a copy of the list, prepared by the Officers, shall, with such alterations as he may see fit to make therein, be delivered by each Member of the Council present and voting, and the names found to have the majority of votes shall form the list to be recommended to the Society.

4. The President and Council shall then nominate by ballot, out of the proposed Council, the persons whom they recommend to the Society for election to the offices of President, Treasurer, Principal Secretaries, and Foreign Secretary, for the ensuing year.

PROCEDURE OF THE COUNCIL IN THE NOMINATION OF FOREIGN MEMBERS.

(*Statutes, Cap. I, Statutes* 19–21.)

19. "A book shall be kept in which Fellows of the Society may enter the names of those men of science whom they suggest as Foreign Members; each entry shall be signed by the proposer, and be accompanied by a statement of the principal grounds on which the suggestion is made, and shall be valid for three years only.

20. "When vacancies are to be filled up, a list of the persons so entered shall be sent to each member of the Council, together with notice of the Meeting at which the list will be considered. At the Meeting thus appointed further entries may be made, and the claims of those men of science whose names have been duly entered in the book shall be considered, and a selection of names shall be made, from among which the Council, at a subsequent Meeting to be then appointed, may make nominations to the Society.

21. "At the second Meeting the selection of the Candidates to be nominated shall be by ballot; when, if two-thirds of the Members of the Council present be in favour of the nomination of any Candidate, his name shall be proposed at the next Ordinary Meeting of the Society, and shall be put to the vote by ballot at the following Ordinary Meeting."

PROCEDURE OF THE COUNCIL IN THE ADJUDICATION OF THE MEDALS.

1. At the first Meeting on the subject of the Medals, the Members of Council are invited to *suggest* a name, or names, which they may deem worthy of consideration in the adjudication of each of the several Medals. The list of suggested names then formed to be entered on the Minutes, with power to Members of Council to add to it afterwards, if they see fit.

2. At a subsequent Meeting (or Meetings), to be held before the Midsummer Recess (at which additions may be made to the List of suggestions), every Member of the Council present is at liberty to *propose* for each Medal the name of a person whom he recommends to be selected to receive it, specifying the particular work or works which form the ground of his recommendation; and these proposals, being seconded, shall be entered on the Minutes. At the same time the proposer is expected to deposit with one of the Secretaries a detailed statement of the claims of the person recommended by him, for consultation by Members of the Council, should they so desire.

3. The Council to be summoned in October, for the purpose of discussing the merits, as regards the award of the Medals, of the persons severally proposed. Additional proposals may be made at this Meeting, if assented to by two-thirds of the Members present.

4. The Council to meet for further consideration of the proposals at an early date thereafter, but before the Anniversary Meeting.

CONDITIONS OF AWARD OF THE ROYAL SOCIETY'S MEDALS.

THE COPLEY MEDAL

is awarded to the living author of such philosophical research, either published or communicated to the Society, as may appear to the Council to be deserving of that honour. The subject or subjects of research, on account of which the medal is awarded, must be specified in making the award.

No limitation is imposed either as to the period of time within which that research was made, or to the particular country to which its author may belong.

The medal may not be awarded to any person who is a Member of the Council at the time when the award is made.

The medal may be given more than once to the same person if the Council deem it expedient.

The medal is, as far as circumstances admit, awarded annually.

THE RUMFORD MEDAL,

consisting of a gold medal with a silver copy struck in the same die, is awarded once every second year "to the author of the most important discovery or useful improvement which shall be made and published by printing or in any way made known to the public in any part of Europe during the preceding two years on Heat or on Light, the preference always being given to such discoveries as, in the opinion of the President and Council of the Royal Society, tend most to promote the good of mankind.

"If during any term of years from the last award no new discovery or improvement shall have been made in any part of Europe relative to Light or Heat, in the opinion of the President and Council of sufficient importance to deserve the award, it may not be given, but the value of it may be reserved, and being laid out in the purchase of additional stock may augment the capital; and the interest of the same, by which the capital may from time to time be so augmented, may be given in money" at a subsequent award with the medal.

THE ROYAL MEDALS,

consisting each of a gold medal with a silver copy struck in the same die, are awarded annually by the Sovereign upon the recommendation of the Council, for the two most important contributions to the advancement of Natural Knowledge, published originally in His Majesty's dominions within a period of not more than ten years, and of not less than one year of the date of the award.

In the award of the Royal Medals one is given in each year to each of the two great divisions of Natural Knowledge.

THE DAVY MEDAL

is awarded annually for the most important discovery in Chemistry made in Europe or Anglo-America.

THE DARWIN MEDAL,

which is accompanied by a grant of £100, is given biennially in reward of work of acknowledged distinction (especially in Biology) in the field in which Mr. Darwin himself laboured. The award may be made either to a British subject or a foreigner, and without distinction of sex.

THE BUCHANAN MEDAL,

which is accompanied by a grant of the balance of the Buchanan Medal Fund which may have accumulated since the last award, is awarded every five years in respect of distinguished services to Hygienic Science or Practice in the direction either of original research or of professional, administrative, or constructive work, without limit of nationality or sex.

THE SYLVESTER MEDAL,

which is accompanied by a grant of the balance of the income of the Sylvester Medal Fund, is awarded triennially for the encouragement of Mathematical Research, irrespective of nationality.

HUGHES MEDAL.

A Gold Medal, bearing a bust of the late Professor D. E. Hughes, is awarded annually, together with the balance of the income of the Fund, to such person as the President and Council may consider the most worthy recipient, without restriction of sex or nationality, as the reward of original discovery in the Physical Sciences, particularly electricity and magnetism or their applications, such discovery or applications having been published not less than one year before the award. If in any year the Council do not see fit to award the medal, owing to no one being deemed sufficiently worthy of it, the income for that year is invested and added to the principal of the Fund.

THE MACKINNON RESEARCH STUDENTSHIPS.

Under the will of the late Sir William Mackinnon the Society has received a bequest to be applied to the foundation and endowment of prizes or scholarships for the purpose of "furthering Natural and Physical Science, including Geology and Astronomy, and of furthering original research and investigation in Pathology," and the following regulations have been drawn up for the administration of the Trust:—

1. Two Studentships, each of the present value of £150 per annum, shall be established under the name of "The Mackinnon Research Studentships."

2. The Awards shall be made by the Council of the Royal Society on the recommendation of a Committee to be appointed by the Council.

3. Each Studentship shall, in every case, be awarded for one year, but, on the recommendation of the Committee, after consideration of a report from the student upon his first year's work, may be awarded to him for a second year. Under exceptional circumstances a Studentship may be renewed for a third year.

4. The Studentships shall be awarded, so far as possible, for investigations in the two main divisions of Science respectively, these divisions corresponding to the two series (A and B) of the 'Philosophical Transactions,' but not including Mathematics.*

5. Applications for the Studentships shall be invited by public advertisement.

6. Candidates shall be required to state whether they hold other endowments, and the Committee shall have power to make inquiry into and take into account the other resources of the candidates.

7. Each candidate shall be required to state the nature of the research in which he proposes to engage. The research for which a Studentship is awarded shall be carried out only at a place approved by the Council, but the student shall not be allowed to carry on other work without the approval of the Council.†

8. The Award shall be made always before the end of the Summer term.

9. In the event of a Studentship not being awarded, or from any cause lapsing before the expiry of the term for which it is granted, the unexpended income of the fund shall be invested so as to be available for extraordinary expenditure in furtherance of the general objects of the Bequest.

10. The Studentships shall be restricted to British subjects.

29 *January*, 1903.

* The following subjects are included in the two main divisions of Science respectively:—

A.		B.	
Astronomy.	Mineralogy.	Anatomy.	Pathology.
Chemistry.	Physics.	Botany.	Physiology.
Geology.		Palæontology.	Zoology.

† The Council do not think it desirable that the Mackinnon Studentship should be held in conjunction with a teaching post.

THE TYNDALL MINING RESEARCH FUND.

In 1910 Mrs. Tyndall, in pursuance of a wish expressed by her husband, the late Professor Tyndall, F.R.S., entrusted to the Society the sum of £1,000, to be administered at the discretion of the President and Council for the purpose of encouraging and furthering research in all matters pertaining to mining, including such questions as ventilation, temperature, diseases incident to miners, and any other lines of scientific enquiry conducive to the improvement of mining and the lot of miners. The fund is invested in £1,010 1*s.* New Zealand 3½ per cent. Stock, and is administered by a Committee of the Society, upon which the Institution of Mining Engineers is represented. Up to the present the income has been utilized for the establishment of an annual Studentship, renewable by re-election, to promote research relating to mining, which would be tenable by suitable persons without imposing restrictions on their ordinary professions or occupations.

REGULATIONS FOR ADMINISTERING THE GUNNING FUND.

A statement of the foundation will be found in the Account of the Society's Trusts, in the 'Record.' The regulations for its administration, proposed by the Council, March 14, 1895, and adopted by the Founder, May 16, 1895, are here subjoined.

Regulations.

1. That the Fund should not be applied in the form of a prize, medal, or reward, but should be devoted to the furtherance of knowledge in some special direction.

2. That, by preference, the interest accruing from the Fund during every three years be applied for the promotion of Physical Science and of Biology alternately.

3. That aid should, by preference, thus be given in Physical Science and Biology respectively, either to investigations or operations which require to be repeated from time to time, or to the development of some specified continued line of research.

In illustration of Regulation 3, the Council suggested as follows:— "Among subjects that would thus seem fitting for the application of the Fund, the following might be given as instances:—The renewal from time to time of magnetic observations in the British Isles; the compilation and publication, at intervals, of detailed lists of well-authenticated spectra; systematic determination of biological data in special regions or under special conditions; assistance to naturalists or others carrying on explorations or special investigations in foreign countries; continued bacteriological observations, similar to those carried out under the direction of the Water Research Committee and others."

REGULATIONS FOR ADMINISTERING THE JOULE FUND.

(*Council Minutes, March* 14, 1893.)

---

1. That the proceeds be applied in the form of a Studentship or Grant, to be awarded every other year, to assist Research, especially among younger men, in those branches of Physical Science more immediately connected with Joule's work.

2. That this Grant be International in its character, and awarded alternately in Great Britain and abroad, or in such order as the President and Council shall from time to time decide.

3. That it be awarded in Great Britain by the President and Council of the Royal Society; and, for award in France, offered to the "Académie des Sciences," Paris; and in Germany, to the "K. Akademie der Wissenschaften," Berlin; or, in any other country, to the leading scientific institution, for award in that country.

4. That the award in Great Britain be made on the recommendation of a Committee, from time to time appointed by the President and Council of the Royal Society, but not of necessity confined to Fellows of the Society.

REGULATIONS FOR THE ADMINISTRATION OF THE GOVERNMENT PUBLICATION GRANT.

(*Council Minutes, June* 15, 1899. *Amended November* 7, 1901.)

The following regulations for the administration of the Publication Grant from H.M. Treasury have been adopted by the Council :—

I. The allotment of the Grant shall be made by the President and Council.

II. In allotting the Grant, the President and Council shall "assist not merely their own publications, but also the adequate publication of scientific matter through other channels and in other ways."

III. In making allotments for the purpose of assisting the adequate publication of scientific matter other than the Society's own publications—

1. The President and Council shall consider—

   (i.) Proposals made by Members of the Council.

   (ii.) Applications made by other Scientific Societies through the usual official channels.

2. Proposals made by Members of the Council may be so made at any meeting of the Council, and applications by other scientific societies shall be reported by the Secretaries to the Council at the first Council Meeting after they have been received; but unless the Council, on grounds of urgency, shall otherwise order, no proposal or application shall be taken into consideration except at the meetings of the Council held in January and July, and no allotment shall be finally decided upon at the first of the said meetings if the decision can conveniently be postponed to the second of the said meetings.

3. Original memoirs shall be considered as having a first claim on the Grant, the aid being given towards the expense either of illustrations or of press-work; but the President and Council shall have power, if they see fit, to make an allotment in aid of other publications which tend to the advancement of natural knowledge, such as reports, abstracts, &c.

4. No decision of the President and Council at any one meeting of the Council, to allot a portion of the Grant, shall be valid unless it receives the support of three-fourths of the members present and voting; but the decision of a simple majority at any one meeting shall be made valid if confirmed by a majority at a subsequent meeting.

IV. The balance of the Grant remaining over at the close of the financial year, after deducting the amounts allotted under Section III, shall be placed to the credit of the General Funds of the Society, to assist in the production of the Society's own publications, unless the President and Council shall otherwise order.

Any Society or individual applying for a Grant to aid the publication of a paper is requested to send in the paper for inspection.

REGULATIONS FOR ADMINISTERING THE SCIENTIFIC RELIEF FUND.*

The history of the Scientific Relief Fund will be found in the account of the Society's Trusts contained in the "Record." The following are the Regulations at present in force:—

REGULATIONS.

1. There shall be a fund called The Scientific Relief Fund, and the object of it shall be to aid such scientific men, or their families, as may from time to time require assistance.
2. All contributions to the fund shall be invested in the name of the Royal Society in such funds as are authorised for investment by Trustees; and in such manner as to form a separate account from that of the Society's other funded property.
3. The fund shall be administered by a Committee, called The Scientific Relief Committee, which shall consist of ten Fellows of the Royal Society, and it shall be the duty of such Committee to select the recipients on whose behalf the income derived from the fund may be properly applied—always reporting thereon to the Council for confirmation.
4. The capital of the fund shall remain entire, and the interest only shall be at the disposal of the Committee.
5. If the whole of the interest shall not be expended in one year, the surplus shall be carried to the next year's account; and, if at any time any surplus in excess of the ordinary income of the year last past shall thus accrue, the Council shall cause the whole, or part of it, to be added to the capital sum already invested; or, should they think fit, may cause any accumulated interest to be invested as unexpended income, the securities purchased being liable from time to time to be realised, and the proceeds expended as income.
6. No application for relief shall be entertained except on the recommendation of the President of one of the following Scientific Societies:—The Chemical, Entomological, Geological, Linnean, London Mathematical, Physical, Royal, Royal Astronomical, Royal Geographical, Royal Meteorological, Royal Irish Academy, Royal Society of Edinburgh, Society of Antiquaries, or Zoological Society.

* Mainly codified from the Original Regulations adopted by the Council Nov. 3, 1859 (see also Minutes of May 26, 1859), and subsequent modifications passed by the Council on Dec. 22, 1859, Jan. 18, 1866, April 30, 1891, Jan. 19, 1893, April 30, 1896, Nov. 5, 1896, May 20, 1909, April 24, 1913.

7. The members of the Committee shall be appointed by the Council, and shall consist of ten members, each of whom shall serve for five years, so that two retire annually, and be not eligible for re-appointment on the occasion of their retiring. Should a vacancy occur by reason of death or otherwise, at any intermediate time, the Council shall appoint a person to fill the vacancy, and the person so appointed shall retire at the time the member whose place he fills would have retired had he continued until then to be a member, but if he have not served more than two years, shall be eligible for re-appointment.
8. The Council shall annually appoint a member of the Committee to act as Chairman for the ensuing year. The Chairman shall have power to nominate one of the Committee to act as his deputy.
9. The Chairman, or his deputy, shall have power to summon a meeting of the Committee at his discretion, and shall fix the time of such meeting.
10. Three of the Committee shall form a quorum.
11. The Treasurer of the Society shall have power, on the requisition of the Chairman of the Committee, or of his deputy, made in pursuance of a resolution of the Committee, but subject, nevertheless, to the provisions of Regulation 12, to make payments out of the Scientific Relief Fund not exceeding £100 in any one case, reporting such action to the Council at its next meeting.
12. The Chairman, or his deputy, shall, notwithstanding Regulation 6, have power to act in urgent cases, after consultation with one of the Secretaries of the Society, without calling the Committee together. In such cases the Chairman shall report his action at the first subsequent meeting of the Committee.
13 If at any time the surplus unexpended income of the Fund mentioned or referred to in the existing Regulation No. 5 (including the value of any securities purchased in accordance with that Regulation, but excluding all income of the then current year) shall amount to more than £700, and at the same time the Committee shall be of opinion that for special reasons a provision of a more permanent character than that contemplated by the existing Regulations ought to be made for the benefit of an applicant for relief, the Committee may (after obtaining the sanction of the Council, and notwithstanding anything to the contrary in the existing Regulations) apply the whole or any part of the excess of such unexpended income over the said sum of £700 in or towards making such provision for such applicant.

In the first Report of the Committee, dated November 30, 1864, it is stated that "It formed no part of the scheme to attempt the grant of annuities; it was rather intended to afford *prompt* relief of the immediate wants of those upon whom sudden affliction had fallen; although at the same time, it in no way debarred a continuation of such relief being given should the funds admit thereof." This intention of the founders, although it has not been embodied in a Regulation, has been continued, as a policy, to the present time.

Applicants are desired to fill in a form which can be obtained from the Assistant Secretary of the Royal Society, in which confidential information is requested upon the following points:—

1. Name, Age, and Social Condition.
2. Nature of Claims, stating scientific work done by the subject of the proposed grant, or by the member of his family on whose scientific claim he relies, appending a list of his principal contributions to science.
3. The nature of the emergency, and how it has arisen.
4. Whether the applicant is receiving, or has received, during the past six months, pecuniary aid from any other source.
5. Whether the applicant is entitled or able, in the circumstances which have arisen, to look to any other assistance; and, if so, what is the source and extent of such expected assistance.
6. Particulars of—

   Number in family.
   How many are self-supporting.
   How many are partially dependent.
   How many are wholly dependent.

7. Are the circumstances of the self-supporting members of the family such as to enable them to render assistance in the case.

In 1886 Sir William (afterwards Lord) Armstrong gave a sum of £7,800 to the Scientific Relief Fund, on the understanding that the said fund should be used for remission of fees in cases of urgent necessity. By a Resolution of Council passed December 10, 1889, "the question of the remission of fees to Fellows of the Society in impecunious circumstances is reserved for the sole consideration of the President and Council of the Society, the amount thus from time to time bestowed being communicated to the Scientific Relief Committee."

THE SORBY RESEARCH FUND.

(*Founded on the terms of the Will of the late Dr. H. C. Sorby, F.R.S.*)

In accordance with the bequest of the Founder, the capital of the Fund (£15,503 16*s*. 6*d*. India $3\frac{1}{2}$ per cent. Stock) remains vested in the President, Council, and Fellows of the Royal Society, to the intent that the income thereof may be devoted to the establishment of a fellowship for the carrying on of original scientific research, the object being to promote the discovery of new facts rather than the teaching of what is known.

In accordance with the desire of the Founder to associate the discovery of new knowledge with his own city of Sheffield, and with his direction that so long as, in the opinion of the Council of the Society, the proposed researches can be properly carried out in connection with the University of Sheffield, and the said University gives the Fellow reasonable facilities in the way of laboratory accommodation and appliances free of charge, then the income shall be administered according to provisions set out in his Will, and embodied in the Rules hereinafter given, and shall be so administered by a Committee consisting of representatives of the Royal Society and of the said University; and in accordance with his suggestion that the Committee should consist of four representatives of the Council of the Royal Society, one of the Council and two of the Senate of the University; and in accordance with his express declaration that it shall be lawful for the Council of the Royal Society to make any rules and regulations which they may think proper with the view of carrying out the provisions of his bequest, and that none of the ideas expressed in the clause of his Will making the bequest shall control the discretion of the parties for the time being making the said rules and regulations, save only that they shall take original scientific research as the fundamental object; and in accordance with his direction that any rules and regulations so made may contain provisions for their subsequent alteration, and his hope that such provisions will be made so as to meet, as far as possible, the exigencies which time and various circumstances educe.

The President and Council of the Royal Society, after consultation with the University of Sheffield, and with the approval of such

University, make the following Rules with regard to the administration of the income of the Fund, which said Rules shall remain in force until revoked or amended by the said President and Council.

RULES.

1. The income of the Fund shall be administered by a Committee consisting of—

(1) Four persons appointed by the Council of the Royal Society;

(2) One person appointed by the Council of the University of Sheffield; and

(3) Two persons appointed by the Senate of the University of Sheffield.

Such appointments shall be made in the month of June or July in each year; and the first meeting of the Committee thus appointed shall be summoned by the President of the Royal Society, who shall fix the time and place of meeting.

2. The members of the Committee shall be appointed for one year only, but shall be eligible for re-appointment in each successive year.

3. The Committee shall choose one of their number to be their Chairman, who shall certify his appointment to the Secretaries of the Royal Society.

4. Meetings of the Committee shall be held in London or Sheffield, as determined by the Committee.

5. Meetings of the Committee shall be summoned by the Chairman when he shall deem it necessary, and he shall also summon a meeting of the Committee whenever required, in writing, to do so by three members of the Committee. The Chairman shall fix the time and place of the meeting.

6. In case the Chairman shall be unable to attend a meeting, the members present shall elect one of their number as Chairman for the time being.

7. Four members of the Committee shall form a quorum.

8. Except as hereinafter expressly provided, every question which comes before the Committee shall be decided by the vote of the majority of the members present and voting. In case of an equality of votes, the Chairman shall have a casting vote in addition to his vote as a member of the Committee.

9. The income of the Fund shall be received in the first instance by the Treasurer of the Royal Society on behalf of the Society, but he shall pay over the same to the Chairman of the Committee, or such other person or persons as the Committee may appoint, on receipt of a requisition certifying that the Committee have so appointed, signed by the Chairman of the meeting at which they so appointed.

10. The Committee shall devote the income of the Fund to the establishment of a fellowship for the carrying on of original scientific research.

11. The Committee may set apart from the yearly income a sum not exceeding fifty pounds to form an apparatus fund. From this fund grants may be made from time to time to the Fellow for the purchase of special apparatus and material required in his research. Such apparatus and material shall remain the property of the University of Sheffield.

12. The Committee may also set apart from the yearly income of the Fund a sum not exceeding five pounds to form a fund to meet the necessary expenses of the Committee: but such expenses are not to include any personal expenses of the members.

13. The balance of the income, after deducting the said sums, shall be paid as a stipend to the Fellow. Any moneys falling in through vacancies or otherwise shall be placed to the account of the apparatus fund. But if the Committee think it advisable they may, after obtaining the consent of the Council of the Royal Society, add such balance or any portion of such accumulated balances to increase the endowment of the fellowship.

14. In making an appointment the Committee shall pay special attention to the capacity for original work of a candidate, as shown by the work already done by him, and to the likelihood that he will continue to do reliable work. The Committee is to bear in mind that the object of the fellowship is not to train men for original research, but to obtain advances in natural knowledge by enabling men of proved ability to devote themselves to research.

15. Each appointment shall be in the first instance for five years, but may in special circumstances be prolonged for further periods if the Committee are satisfied with the Fellow's work.

16. An annual report of the work done by the Fellow shall be presented to the Committee. Should the Committee find that the Fellow does not give attention to his duties, it shall have the power

to remove him from his fellowship, but it shall not do so merely because his work has produced no definite results.

17. A Fellow may be removed from his fellowship if he shall do any act which may, in the unanimous opinion of the members of the Committee present at the meeting at which such act is considered, be detrimental to the interests of the University of Sheffield. Provided always, that such removal shall forthwith be notified to the Council of the Royal Society, and shall not take effect unless and until it receives the approval of the Council.

18. No Fellow shall be removed except after at least fourteen days' notice in writing has been given to him of the intention to remove him, which notice shall specify the cause of the proposed removal, and shall fix a day, subsequent to the expiration of the fourteen days, on which the Committee will meet and hear the Fellow in his own defence.

19. A new appointment shall preferably be in a subject different from the previous one.

20. No Fellow shall be permitted to undertake any permanent duties outside his research work.

21. The research shall be carried out in one of the laboratories of the University of Sheffield. This condition may, however, be dispensed with when the nature of the investigation requires that the work should be done elsewhere.

22. These Rules may be revoked or amended by the Council of the Royal Society at any time after giving the University and the Committee an opportunity of stating their views on th proposed changes.

THE TRAVELLING EXPENSES FUND.

This Fund has been founded by Professor Arthur Schuster with a donation of £1,500 *for the purposes set forth in the following scheme.*

1. The Capital of the Fund shall consist of the sum contributed by the founder, Professor Arthur Schuster, together with such additions as may be made thereto from time to time by voluntary contributions.

2. The Capital shall be invested, in the name of the Royal Society, in such securities as the President and Council may from time to time determine in their absolute discretion.

3. The income shall be used to pay the travelling expenses of the delegates sent by the Royal Society to the meetings of the International Association of Academies, subject to the following conditions:—

   (1) Not more than four delegates to the General Assembly, of whom not more than two shall be officers of the Society, shall benefit from the Fund.

   (2) One delegate only to the meetings of the Council, who may or may not be an officer of the Society, shall benefit from the Fund.

   (3) First-class return fares only shall be provided out of the Fund, unless a special grant is made under 4 (*c*).

4. When a sufficient balance has accumulated, the surplus may be used for any of the following objects at the discretion of the President and Council of the Royal Society:—

   (*a*) To pay the travelling expenses of delegates sent by the Royal Society to international meetings, other than those of the International Association of Academies.

   (*b*) To pay the travelling expenses of more than four delegates to the General Assembly of the International Association of Academies.

   (*c*) To provide a sum towards maintenance of the delegates sent by the Royal Society to the meetings of the International Association of Academies, or other international meetings, such sum not to exceed £1 1*s*. 0*d*. per day during the time of necessary absence from England.

5. In making grants under (*a*), (*b*), and (*c*), the Council of the Royal Society will bear in mind the fact that as, occasionally, the International Association of Academies may meet in distant countries, a substantial balance should always be available for such contingency.

6. In the event of the International Association of Academies ceasing to exist, the income of the fund shall be used to pay the travelling and maintenance expenses of delegates sent by the Royal Society to other international meetings, at the discretion of the President and Council.

[The term "officer of the Society" shall be understood to include the President, Treasurer, and Secretaries, but not the Vice-Presidents.]

NATIONAL PHYSICAL LABORATORY.

SCHEME OF ORGANIZATION.

1. The name of the Institution shall be the National Physical Laboratory. The Kew Observatory shall be incorporated therewith.

2. The ultimate control of the Institution shall be vested in the President and Council of the Royal Society, who in the exercise thereof may from time to time issue such directions as they may think fit to the General Board and Executive Committee hereinafter described. The President of the Royal Society shall be the Chairman of the Governing Body as hereinafter defined. The income and all other property of the Institution shall be vested in the Royal Society for the purposes of the Institution.

3. For the present, and until otherwise ordered by the President and Council of the Royal Society, with the approval of H.M. Treasury, there shall be a Governing Body for the Institution, consisting of a General Board and an Executive Committee, the constitution and duties of which shall be as hereinafter defined. Provided always that the Permanent Secretary of H.M. Board of Trade shall be *ex officio* a member of the Governing Body, and that the choice of members of the Governing Body, or of any Committee thereof shall not be confined to Fellows of the Royal Society.

4. The General Board shall consist of the President, Treasurer, and Secretaries of the Royal Society, the Vice-Chairman of the Board (appointed as defined below by the President and Council of the Royal Society), the Permanent Secretary of the Board of Trade, and of thirty-six ordinary members.

Twenty-four of the ordinary members shall be appointed by the President and Council of the Royal Society; of the remaining twelve ordinary members, two shall be nominated for appointment by the Council of each of the following Institutions, as being fitted to represent commercial interests in connection with the Laboratory:—

The Institution of Civil Engineers.
The Institution of Mechanical Engineers.
The Institution of Electrical Engineers.
The Iron and Steel Institute.
The Institution of Naval Architects.
The Society of Chemical Industry.

In the selection of ordinary members of the General Board care shall be taken that Scotland and Ireland are represented.

Any person not being already a member of the General Board who shall become a member of the Executive Committee, shall be a member of that Board during his tenure of office on the Executive Committee, but shall be regarded as an additional, and not as an ordinary, member of the Board.

5. The Executive Committee shall consist of the President, Treasurer, and one of the Secretaries of the Royal Society; the Vice-Chairman of the Executive Committee (appointed as defined below); the Permanent Secretary of the Board of Trade; six persons appointed by the President and Council of the Royal Society from among those who are members of the Kew Observatory Committee at the time when the Kew Observatory is incorporated in the National Physical Laboratory (two of these six persons shall retire at the end of every two years, and vacancies occurring amongst them by retirement or otherwise shall not be filled up); and of twelve ordinary members.

The ordinary members shall be nominated by the President and Council of the Royal Society, but one-half shall be chosen from among those members of the General Board who have been nominated as fitted to represent commercial interests on that Board.

*Those members of the Executive Committee who are Fellows of the Royal Society, shall be appointed by the President and Council to be the Gassiot Committee of the Royal Society.

6. The Vice-Chairman of the General Board shall be appointed by the President and Council of the Royal Society, and shall also be Vice-Chairman of the Executive Committee. He shall hold office for six years, and shall be eligible for re-appointment, but shall not hold office for more than twelve years.

7. At least one-sixth of the ordinary members of the General Board and of the Executive Committee shall retire annually.

In the case of the General Board, the retiring ordinary members shall be selected by seniority, four being selected from the members nominated by the President and Council of the Royal Society, and two from the members nominated by the Technical Societies named in the scheme.

In the case of the Executive Committee, the retiring ordinary members shall be selected by seniority, one being selected from the members nominated by the President and Council of the Royal Society, and one from the members nominated by the Technical Societies named in the scheme.

* This clause ceased to be operative in 1910. (*See* p. 127.)

No retiring member of the General Board or of the Executive Committee shall be eligible for re-appointment until at least one year has elapsed from the date of his retirement.*

The President and Council shall have power to remove from the General Board and from the Executive Committee any member of either whom they may judge to be disqualified.

Vacancies on the General Board or on the Executive Committee due to death, resignation, or removal by the President and Council of the Royal Society, shall be filled by the President and Council of the Royal Society, provided always that—

(1) Any person so appointed shall, for the purposes of the regulations for retirement from the Board or Committee, be regarded at the time of his appointment as having served for the same period as the member to whose place he succeeds.

(2) If the vacancy on the General Board be caused by one of the persons nominated as fitted to represent commercial interests ceasing to be a member of the Board, the President and Council of the Royal Society shall choose his successor from among a list of names recommended by the Councils of the Institutions named in Section 4.

(3) If a vacancy on the Executive Committee be caused by one of the persons nominated as fitted to represent commercial interests ceasing to be a member of the Committee, his successor shall either be selected from among those members of the General Board who were nominated as fitted to represent commercial interests, or shall be nominated by the President and Council of the Royal Society after consultation with the Councils of the Institutions named in Section 4.

The President and Council of the Royal Society shall determine the order of the seniority of the members of the first General Board and of the first Executive Committee for the purposes of the regulations for retirement.

The Executive Committee.

8. The Executive Committee shall have the immediate management of the National Physical Laboratory; shall appoint and dismiss the officials, except the Director; and shall determine the nature of the work to be undertaken from time to time.

* The following addition to this paragraph, proposed by the Royal Society, has been sanctioned by H.M. Government:—

"Unless, in the opinion of the President and Council of the Royal Society, there are special reasons in the case of any retiring member for his services being retained."

The General Board.

9. A meeting of the General Board shall be held in June, at which the Executive Committee shall present a report on the work and finances of the National Physical Laboratory during the year ending on the preceding March 31. Copies of this report shall be circulated among the members of the General Board at least one week before the meeting, and after the meeting shall be forwarded to the President and Council of the Royal Society, together with any further report, resolutions, or recommendations which may be added by the General Board.

The Executive Committee shall also lay before the General Board at its meeting in June a statement as to the work which it is proposed to undertake in the Laboratory during the ensuing year. This statement shall be circulated among members of the Board at least a week before the meeting; and the General Board may make such recommendations relative to the statement, or to the future work of the National Physical Laboratory, as they may think fit.

These recommendations shall be laid before the Executive Committee for their consideration.

Sub-Committees.

10. The Executive Committee may from time to time appoint Sub-Committees, of which the members shall not necessarily be members of the Executive Committee or of the General Board, either to superintend or to assist in certain specified investigations, or to superintend some department of the National Physical Laboratory.

The Director.

11. The Director of the National Physical Laboratory shall be appointed by the President and Council of the Royal Society after consultation with the Executive Committee, on such terms as the President and Council may determine, and shall be removable by the President and Council. He shall be responsible to, and shall take instructions from, the Executive Committee, but, subject to such instructions, he shall have the sole direction and control of the officials of the National Physical Laboratory and of the work done within it.

The Executive Committee may delegate its power of appointing and dismissing the officials of the Institution to the Director in such cases as it may think fit.

The Director shall neither be allowed nor be called upon to undertake work not connected with the National Physical Laboratory, except with the consent of the Executive Committee.

Finance.

12. The Royal Society shall open a banking account, to be called "The National Physical Laboratory Account of the Royal Society," into which all sums received by the Executive Committee for the purposes of the Institution shall be paid. The Treasurer of the Royal Society shall also pay into this account all sums received by him for the said purposes, after deducting therefrom such amounts as he shall be directed by the President and Council, with the approval of the Treasury, to retain for the purpose of defraying any expenses which the Royal Society may incur in the exercise of its control of the Institution.

The Executive Committee shall be empowered to draw on this account for the purposes of the Institution by cheques signed by such members of the Executive Committee as may be authorised by the Committee to do so.

Legal Proceedings.

13. Any legal proceedings with regard to the affairs of the Institution, which it may become necessary to institute or defend, shall be instituted or defended by the Solicitors of the Royal Society, in the name and on behalf of the Royal Society upon the instructions of the Executive Committee, but no such proceedings shall be instituted or defended without the order of the President and Council of the Royal Society.

The Kew Observatory Committee of the Royal Society.

"The Kew Observatory Committee of the Royal Society," incorporated under the Companies Act, 1867, shall be wound up; and the property thereof shall be held by the Royal Society for the purposes of the Institution.*

GENERAL BOARD OF THE NATIONAL PHYSICAL LABORATORY.

Ex-Officio
Members.

The President of the Royal Society.
The Vice-Chairman of the Board (Lord Rayleigh, O.M., F.R.S.).
The Treasurer of the Royal Society.
The Secretaries of the Royal Society.
The Permanent Secretary of the Board of Trade.

* This paragraph is superseded by the new Scheme for Kew, Eskdalemuir, and Valencia Observatories. (*See* p. 127.)

Retires March 31		
1917	Mr. Dugald Clerk, F.R.S.	Nominated by the President and Council of the Royal Society.
1917	Prof. B. Hopkinson, F.R.S.	
1917	Prof. J. E. Petavel, F.R.S.	
1917	Mr. Otto Beit	
1918	Col. R. E. Crompton	
1918	Mr. R. L. Mond	
1918	Major W. A. J. O'Meara	
1918	Hon. Sir Charles A. Parsons, F.R.S.	
1919	Sir H. F. Donaldson	
1919	Viscount Esher	
1919	Sir Robert Hadfield, F.R.S.	
1919	Sir J. J. Thomson, O.M., F.R.S.	
1920	Dr. E. H. Griffiths, F.R.S.	
1920	Vice-Adml. Sir Henry Jackson, F.R.S.	
1920	Mr. Leslie Robertson	
1920	Sir J. Wolfe Barry, F.R.S.	
1921	Dr. G. T. Beilby, F.R.S.	
1921	Sir Alfred Ewing, F.R.S.	
1921	Sir Francis Hopwood	
1921	Lord Moulton, F.R.S.	
1922	Mr. Horace Brown, F.R.S.	
1922	Mr. J. M. Falkner	
1922	Prof. C. H. Lees, F.R.S.	
1922	Admiral Sir Henry Oram, F.R.S.	
*1917	Mr. W. Duddell, F.R.S.	

1917	Mr. W. H. Ellis (Iron and Steel Inst.)	Nominated by the Technical Societies named in the Scheme.
1917	Prof. W. C. Unwin, F.R.S. (Inst. Civil Eng.)	
1918	Sir J. F. C. Snell (Inst. Elect. Eng.)	
1918	Mr. T. Tyrer (Soc. Chem. Industry)	
1919	Mr. R. Elliott Cooper (Inst. Civil Eng.)	
1919	Sir Hugh Bell, Bart. (Iron and Steel Institute)	
1920	Sir William E. Smith (Inst. Naval Arch.)	
1920	Mr. M. Longridge (Inst. Mech. Eng.)	
1921	Mr. N. H. Martin (Soc. Chem. Industry)	
1921	Mr. R. T. Smith (Inst. Elect. Eng.)	
1922	Sir A. F. Yarrow (Inst. Naval Arch.)	
1922	Capt. H. R. Sankey (Inst. Mech. Eng.)	
*1918	Mr. Alexander Siemens (Inst. Elect. Eng.).	

* Additional Members.

EXECUTIVE COMMITTEE OF THE NATIONAL PHYSICAL LABORATORY.

Retires March 31		
	The President of the Royal Society.	
	The Vice-Chairman of the Committee (Lord Rayleigh, O.M., F.R.S.).	
	The Treasurer of the Royal Society.	
	A Secretary of the Royal Society (Prof. A. Schuster).	
	The Permanent Secretary of the Board of Trade.	
1917	Mr. W. Duddell, F.R.S.	
1918	Col. R. E. Crompton.	
1919	Sir Robert Hadfield, F.R.S.	
1920	Dr. E. H. Grffiths, F.R.S.	
1921	Sir Francis Hopwood.	
1922	Prof. C. H. Lees, F.R.S.	
1922	Sir H. F. Donaldson.	
1917	Mr. W. H. Ellis	Members of the General Board nominated by the Technical Societies named in the Scheme.
1918	Mr. Alexander Siemens	
1919	Prof. W. C. Unwin, F.R.S.	
1920	Sir William E. Smith	
1921	Mr. T. Tyrer	

Director.
R. T. Glazebrook, C.B., D.Sc., F.R.S.

Superintendent of Engineering Department.
T. E. Stanton, D.Sc., F.R.S.

Superintendent of the Department of Metallurgy and Metallurgical Chemistry.
W. Rosenhain, D.Sc., F.R.S.

Superintendent of The William Froude National Experimental Tank.
G. S. Baker.

REGULATIONS FOR ADMINISTERING THE GOVERNMENT GRANT FOR SCIENTIFIC INVESTIGATIONS.

I.

1. The Government "Grant for Scientific Investigations undertaken with the sanction of a Committee appointed for the purpose" shall be administered by the President and Council of the Royal Society; but, subject as hereinafter provided, no payment shall be made thereout except with the sanction of a General Committee, consisting of the President and Council of the Royal Society for the time being, of the following *ex officio* Members:—

The President of the Royal Society of Edinburgh and one other Representative,
The President of the Royal Irish Academy and one other Representative,
The Presidents of—
The British Association,
The London Mathematical Society,
The Royal Astronomical Society,
The Physical Society,
The Institution of Civil Engineers,
The Institution of Mechanical Engineers,
The Institution of Electrical Engineers,
The Chemical Society,
The Iron and Steel Institute,
The Geological Society,
The Royal Geographical Society,
The Linnean Society,
The Zoological Society,
The Royal Anthropological Institute,
The Royal College of Physicians,
The Royal College of Surgeons,

and of the Members, for the time being, of the several Boards hereinafter spoken of.

2. The President of the Royal Society, or, in his absence, one of the Vice-Presidents, shall preside at all Meetings of the General Committee.

3. Seven Boards shall be established, viz.:—

A. For the consideration of Applications relating to Mathematics. Mathematical Physics, Crystallography, and Mathematical Astronomy.

B. For the consideration of Applications relating to Experimental Physics, Observational Astronomy, and Meteorology.

C. For the consideration of Applications relating to Chemistry and Metallurgy.

D. For the consideration of Applications relating to Geology, Palæontology, Mineralogy, and Geography.

E. For the consideration of Applications relating to Botany.

F. For the consideration of Applications relating to Zoology and Comparative Anatomy.

G. For the consideration of Applications relating to Animal Physiology and Medical Subjects.

4. Each Board shall consist of eight Members, to be appointed by the President and Council of the Royal Society, Scotland and Ireland being as far as possible represented on each Board, and each Member shall serve for four years, so that two retire annually, and be not eligible for re-appointment on the occasion of their retiring. Should a vacancy occur by reason of death or otherwise, at any intermediate time, the Council shall appoint a person to fill the vacancy, and the person so appointed shall retire at the time the Member whose place he fills would have retired had he continued until then to be a Member, but if he have not served more than two years shall be eligible for re-appointment.

5. The President and Council of the Royal Society shall appoint a Member of each Board to be Chairman of the Board. All communications made to and by the Board shall be made through the Chairman, who shall be held responsible for the management of the business of the Board, and who shall have a second or casting vote. When a Chairman is unable to perform the duties of the Chair, he shall appoint a Member of the Board to act as his deputy, and to exercise his powers.

6. Each Chairman has authority to summon his Board, whenever he thinks fit (in addition to any Meeting or Meetings of the Board which may be appointed by the Council), to meet either at the Rooms of the Royal Society, during the hours specified in the Statutes, or at such other place as he may deem desirable.

7. The summonses are to be issued by the Clerk at the direction of the Chairman.

8. Any three Members of a Board are to be a quorum of that Board; but the decisions arrived at at a Meeting of a Board at which less than three Members are present shall be valid, if subsequently agreed to in writing by not less than five Members in all.

II.

9. In order to meet any extraordinary demands which may be made upon the Grant, a Reserve Fund may be accumulated, but so that it shall not at any time exceed £2,000.

10. A Grant, the payment of which is intended to be completed within a period of two years following upon the Meeting of the Committee at which the Grant was made, shall be called an "ordinary" Grant. The Committee shall, however, if they see fit, make Grants for "personal" or other expenditure, each of which may extend over a period not exceeding three years, but in no case shall such a personal Grant exceed £300 per annum. For this purpose the Committee may, in any one year, reserve from the Fund of the year an amount sufficient to cover the payment during the period for which the Grant has been made, the continuance of the payment of the instalments of such Grants to be conditional on the recipients furnishing, as hereinafter provided, evidence satisfactory to the Committee that the object of the Grant is being properly carried out. Such Grants shall be called "extended" Grants.

III.

11. Adequate notice shall be given in the public papers each year that applications for Grants must be sent in to the Royal Society not later than the first day of January, and no applications received after that date shall be considered by the Committee of that year.

12. Each applicant shall be required to furnish information under the following heads:—

a. The nature of the research in which he desires to engage, and of the scientific results expected to follow therefrom.

b. The amount asked for, and particulars of the proposed expenditure.

c. Whether he has received any previous Grant from any source for the same object, and if so, with what results.

d. Whether any portion of the Grant is to be devoted to his own personal expenses.

e. What apparatus, if any, of permanent value he will require; so that any instruments, already at the disposal of the Committee, may be utilised.

13. Applications which have for their avowed object the mere promotion of commercial inventions or patents, and not any scientific research, and applications which appear to the Secretaries of the Royal Society to have reference to matters wholly unsuitable for Grants, shall not be included in the list referred to in Regulation 14.

14. As soon as possible after January 1st in each year, the Secretaries of the Royal Society shall cause to be drawn up a list of all the applications, arranged, according to the nature of the research in each application, in classes corresponding to the above-mentioned Boards, and shall cause such list to be distributed to all Members of the Committee. This list shall contain a brief statement of the information received under Clause 12.

15. The Secretaries of the Royal Society shall further cause to be sent to the Chairman of each Board the original applications belonging to the class corresponding to his Board, together with any other information, letters, documents, &c., which may have been furnished by the several applicants.

16. If the Chairman of a Board, on receiving a list of applications under Regulation 15, shall find that any application on that list is, in his opinion, more appropriate to another Board than his own, or that any application which ought, from its nature, to have been referred to a Board or to Boards besides his own, is referred only to his own Board, or that an application proper to his Board has been referred to another Board, he shall at once report the same to the Secretaries of the Royal Society.

17. Should any application appear to the Secretaries of the Royal Society to relate to more than one Board, they shall refer the application to the several Boards to which it appears to relate.

18. Each Board, having taken into consideration the applications submitted to it, making such use of correspondence between Members of the Board as may be desirable for the purpose, shall report, stating, with reference to each such application, whether they recommend the acceptance of it in part or in whole, or the rejection of it; and the Clerk shall cause the Reports of the several Boards to be distributed as soon as possible to all Members of the Committee.

19. It shall be in the power of any Board to initiate an inquiry and to recommend a Grant for the purpose, and such a recommendation

having been reported to the Committee with the other recommendations of the Board, shall take its place among applications recommended to the Committee for acceptance, in spite of application not having been made in the ordinary way.

20. In the ordinary course in each year a Meeting of each Board shall be held at the Society's Rooms soon after the receipt by the Chairman of the applications, and a subsequent Meeting shall be held to come to final decisions on the applications on a day to be fixed by the Council for a simultaneous Meeting of the Boards; but the Chairman may, if he finds it desirable, change the day of the latter Meeting, and he may even omit the one or the other of these Meetings, should he judge the one or the other to be unnecessary.

IV.

21. The meeting of the General Committee shall be held at as early a date as possible after the simultaneous meeting of the Boards, as may be determined by the President and Council in each year, at which Meeting the Reports of the Boards shall be read, considered (the Chairman of each Board, or in his place some other Member of it, giving such explanations with regard to the decisions of the Board as may seem desirable), and voted upon. The voting shall be by show of hands, unless any Member demands a ballot, in which case it shall be by ballot.

22. In the case of applications which have been recommended by the appropriate Board, or recommendations initiated by any Board, the voting in Committee shall be by simple majority of those present, except in the case of "extended" Grants coming under Clause 10, which Grants shall require the assent of two-thirds of those present.

23. Applications which have been rejected by the appropriate Board shall not be reconsidered in Committee except with the consent of two-thirds of those present, and any applications so reconsidered shall not be granted by the Committee otherwise than by a majority of two-thirds; likewise a proposal to increase the amount of any Grant made by a Board shall not be considered in Committee except with the consent of two-thirds of those present, and the increase so considered shall not be granted by the Committee otherwise than by a majority of two-thirds.

24. The Committee shall have power to place each year at the disposal of the President and Council of the Royal Society, a sum not exceeding £500 to meet any pressing demands upon the Fund which may be made between the annual Meetings of the Committee.

25. The President of the Royal Society shall further have power, in case he is of opinion that there is urgency for an immediate Grant of a sum too large to be provided by the Fund referred to in Clause 24, and necessitating a call upon the Reserve Fund, to summon a Special Meeting of the Committee, who, if they see fit, shall decide on such Grant, provided always that due notice of such Meeting, with a statement of the purpose for which it is called, be sent to each Member of the Committee fifteen days before the date fixed for the Meeting.

V.

26. No Grants shall be made to other than British subjects.

27. All Grants shall be subject to the following conditions, and every applicant shall, on his applying, be duly informed of these conditions:—

i. That all instruments, specimens, objects, or materials of permanent value, whether purchased or obtained out of, or by means of, the Grant, or supplied from among those at the disposal of the Committee, are to be regarded, unless the Committee decide otherwise, as the property of the Government, and are to be returned by the applicant, for disposal according to the orders of the Committee, at the conclusion of his Research, or at such other time as the Committee may determine.

ii. That every one receiving a Grant shall furnish to the Committee, on or before the 1st of January following upon the allotment of the Grant, a Report (or, if the object of the Grant be not then attained, an interim Report, to be renewed at the same date in each subsequent year until a final Report can be furnished, or the Committee dispense with further Reports), containing (*a*) a brief statement showing the results arrived at, or the stage which the inquiry has reached; (*b*) a general statement of the expenditure incurred, accompanied, so far as is possible, with vouchers; (*c*) a list of the instruments, specimens, objects or materials, purchased or obtained out of the Grant, or supplied by the Committee, which are at present in his possession; and (*d*) references to any Transactions, Journals, or other publications in which results of the Research have been printed.

In the event of the grantee failing to send in, within three months of the said 1st January, a report satisfactory to the Board which recommended the Grant, he shall return the whole of the sum allotted to him.

iii. That when a Grant is asked for a definite Research, for which an estimate can be obtained, applicants are required, with their applications, to furnish such an estimate.

iv. That no portion of a Grant may be expended in the payment of an assistant unless specific application for the sanction of such expenditure is included in the statement required to be furnished under Regulation 12*b*.

v. That when an application is for a Grant to two or more persons to act as a Committee for the purpose of carrying out some scientific object, the application shall state which Member of the proposed Committee is willing to act as Secretary, to be responsible for furnishing the Report, for receiving and disbursing the money, and in general for the conduct of the business of the Committee.

vi. That Grants shall lapse at the end of two years from the date of allotment, if application for payment be not made within that time.

vii. That papers in which results are published which have been obtained through and furnished by the Government Grant, should contain an acknowledgment of that fact.

The Committee shall further have power to attach to any Grant any other conditions which they may think desirable.

28. Every applicant to whom a Grant is made shall, before any of the Grant is paid to him, be required to sign an engagement that he is prepared to carry out the general conditions applicable to all Grants, as well as any conditions which may be attached to his particular Grant.

29. The Chairman of each Board is expected to see that the Annual Reports provided for by Regulation 27 (ii), furnished by Grantees, give an adequate account of the work done and the results attained, and in cases where the Reports are inadequate, to inform the Clerk of the fact in order that he may communicate with such Grantees.

30. Printed copies of these Reports shall each year, so soon as possible after January 1, be submitted to the several Boards; and it shall be the duty of each Board to examine the Reports relating to Grants recommended by it, and to report to the Committee (or, in case of urgency, to the Council of the Royal Society) any deficiencies therein, or any action relating thereto which the Board thinks desirable.

31. The Clerk to the Committee shall report to the several Boards all cases in which Grantees refuse or neglect to furnish Reports, and it shall be the duty of each Board to whom any such case is reported to take such action thereon as it may think necessary, including a recommendation to the Council that legal proceedings should be taken against the defaulting Grantee.

32. In the case of a Grant recommended by a Board being for the purpose of enabling the applicant to collect by means of the Grant, or part of it, specimens, objects, or materials of permanent value, the Board shall, whenever it is able to do so, add to its recommendation conditions as to the final disposal of such specimens, objects, or materials.

33. The recipient of an "extended" Grant shall make to the Board which recommended the Grant, half-yearly, or, if the Board desire it, oftener, such Reports as the Board may determine concerning the way in which the object of the Grant is being carried out; and each such recipient shall, on receiving notice that the Grant has been made to him, be informed of his duty to make such Reports, and shall express in writing his willingness to do so. Should any Board be of opinion, after receiving such Reports, that the object of the Grant is not being properly carried out, they shall report the same to the next Meeting of the Committee. The Chairman of the Board shall move at the Meeting of the Committee that the Grant be discontinued, and if the Committee by a majority approve of the Grant being discontinued, it shall be discontinued.

VI.

34. The President and Council of the Royal Society may in each year set aside out of the Reserve Fund such sum as they may consider desirable to provide for any expenditure which may be incurred by the Royal Society (including expenditure on printing, clerks' salaries. and office expenses) in undertaking, controlling, supervising or advising upon matters which the President and Council may at the request of the Government undertake, control, supervise, or advise upon.

VII.

35. There shall be a Clerk to the Government Grant Committee, who shall be appointed by the President and Council of the Royal Society, and shall be paid such salary as they shall determine. He shall be under the orders and directions of the President and Council of the Royal Society.

36. A sum of not exceeding £250 shall be yearly placed at the disposal of the President and Council to provide for the salary of the Clerk, and other incidental purposes.

37. All Grants and other payments made out of the Government Grant shall be paid by cheque, signed by the Treasurer of the Royal Society and countersigned by the Clerk to the Committee.

38. A Professional Accountant shall be employed to audit the accounts in chief. A preliminary examination of the detailed accounts and vouchers shall be made by the Clerk to the Committee, who is instructed to submit to the Chairman of the appropriate Boards the cases which appear to him to be doubtful; and the Chairman of a Board shall examine any such case so submitted to him, and take such action as may seem to him desirable.

VIII.

39. A Schedule shall be kept of all instruments, specimens, &c., of permanent value, in furtherance of Regulation 27 (i), and of Clause *e* of Regulation 12.

40. The Chairman of a Board may authorise the transfer of any instrument, specimen, &c., obtained by means of a Government Grant, and no longer needed by the person by whom it was obtained or to whom it was assigned, to any other person applying to the Government Grant Committee for the loan of the instrument, specimen, &c., if in his judgment such a transfer is desirable. He shall in each case report his having done so to the Secretaries of the Royal Society.

41. All instruments returned by a Grantee which are not suitable for preservation under Regulation 39, may be purchased by such Grantee for such price as may be approved of by the Treasurer of the Royal Society, and if not so purchased, shall be offered for presentation to the National Physical Laboratory, and if the Laboratory has no need for them, they shall be sold to such person and for such price as may be approved by the Treasurer of the Royal Society. The proceeds of any such sale shall be carried into the Reserve Fund.

June, 1904. Amended *April* 30, 1908; *June* 23, 1910.

Account of the Appropriation of the Government Grant annually voted by Parliament for "Scientific Investigations undertaken with the sanction of a Committee appointed for the purpose."

For the Year 1915–16.

	£	*s.*	*d.*
S. Chapman, lunar diurnal variation of terrestrial magnetism	60	0	0
R. J. Strutt, electric discharge investigations	108	0	0
C. Chree, phenomena presented by magnetic daily range over long series of years	50	0	0
J. E. Purvis, ultra-red absorption spectra of the vapours and solutions of organic substances	80	0	0
O. W. Richardson, kinetic energy of electrons	60	0	0
T. M. Lowry, (*a*) optical rotatory dispersion; (*b*) rotatory dispersive power of organic compounds	60	0	0
H. E. Armstrong, (*a*) action of the polybasic acids; (*b*) synthetic products of the action of some of the enzymes	125	0	0
A. Lapworth, investigation of the pungent principle of ginger and of allied plant products	20	0	0
M. Nierenstein, attempts to synthesise anthocyanines	5	0	0
G. Senter, (*a*) kinetic investigations on the Walden Inversion; (*b*) viscosity in binary mixtures of liquids	25	0	0
K. J. P. Orton, influence of substituents on the speed of chlorination of benzene and naphthalene derivatives	30	0	0
J. H. T. Roberts, investigation into the nature of high temperature chemical reactions	5	0	0
W. H. Mills, investigation of molecularly asymmetric compounds	15	0	0
W. E. S. Turner, nature of molecular association	10	0	0
T. S. Moore, asymmetric tervalent nitrogen	5	0	0
F. S. Kipping, organic derivatives of silicon and of tin	25	0	0
G. T. Morgan, (*a*) co-ordination compounds of the rarer metals; (*b*) organic derivatives of mercury, selenium, and allied elements; (*c*) constitution of diazo compounds	20	0	0
Carried forward	£703	0	0

Brought forward...........	£703	0	0
R. H. Pickard, relation between optical activity of organic compounds and their constitution	90	0	0
F. C. Thompson, colour of metals	10	0	0
W. H. Perkin, (*a*) constitution of cryptopin, and allied alkaloids; (*b*) constitution of harmine and harmaline; (*c*) derivatives of epicamphor	50	0	0
M. S. Leslie, determination of the molecular weights of thorium and actinium emanations	5	0	0
R. H. A. Plimmer, natural and synthetical organic phosphorus compounds; analysis of proteins, and some metabolism experiments	25	0	0
C. Reid, fossil characeæ	20	0	0
R. R. Marett, excavation of a palæolithic site in Jersey...	50	0	0
G. K. Sutherland, study of marine fungi	10	0	0
R. P. Gregory, experimental study of heredity	65	0	0
M. T. Fraser, distribution of phytoserol, &c., in plants and animals	15	0	0
R. Southern, an ecological survey of Malahide Inlet, Co. Dublin	20	0	0
W. J. Dakin, an expedition to investigate the flora and fauna of the Abrolhos Islands	80	0	0
J. H. Orton, rate of growth and age of sex-change in *crepidula*	12	0	0
N. M. Haldane, colour inheritance in mice...............	5	0	0
J. H. Lloyd, investigations on a trematode worm parasite in dogfish..	5	0	0
G. Elliot Smith, anatomy of the brain of ceratodus ...	12	0	0
Karl Pearson, heredity in dogs	50	0	0
R. C. Punnett, heredity of *heterochromia iridis* in rabbits	50	0	0
F. M. Durham, physiology of inheritance	15	0	0
R. Assheton, development of *dolichoglossus serpentinus*	11	0	0
C. J. Patten, migration routes of birds	10	0	0
W. H. Harvey, presence and physiological function of smooth muscle tissue normally present in the kidney ...	30	0	0
Carried forward...........	£1,343	0	0

	£	s.	d.
Brought forward............	£1,343	0	0
R. G. Canti, occurrence of urea in cerebo-spinal fluid	10	0	0
P. Hamill, disturbances of circulation in relation to arterial changes	15	0	0
M. S. Pembrey, metabolism under various conditions of health and disease............	30	0	0
D. Waterston, development of the heart in man	15	0	0
A. R. Cushny, investigations on (*a*) respiratory centre; (*b*) antagonism of poisons	20	0	0
F. A. Bainbridge, factors concerned in the adjustment of the circulatory mechanism for the influence of gravity	10	0	0
E. H. Starling, performance of the mammalian heart	60	0	0
W. D. Halliburton, physiology of cerebro-spinal fluid and allied problems	50	0	0
F. H. Thiele, further investigation into the problem of immunity and infection	30	0	0
D. Thacker, influence of electrolytes on living tissues	20	0	0
C. R. A. Thacker, relation of electrolytes to living tissues............	20	0	0
C. L. Evans, relation of gaseous metabolism and energy expenditure to mechanical performance in the mammalian heart under various conditions	30	0	0
T. Lewis, relation of hypertrophy to cardiac valve lesions	25	0	0
J. Barcroft, physical and chemical properties of hæmoglobin in pure solutions and in blood, &c.	40	0	0
S. W. Cole, effect of pancreatin on post-anæsthetic acidosis	40	0	0
A. E. Boycott, experimental nephritis............	30	0	0
R. L. M. Wallis, application of the study of protective ferments in the blood to the diagnosis of pancreatic disease and the possible relationship to diabetes mellitus	20	0	0
G. Graham, variations in the blood sugar	20	0	0
C. S. Sherrington, enquiry into the action of tetanus toxin on the nervous system	30	0	0
W. E. Dixon, investigations on the cerebro-spinal fluid	10	0	0
Carried forward............	£1,867	0	0

Brought forward	£1,867	0	0
A. F. S. Kent, researches on the heart	10	0	0
H. Chambers, pathology of Graves' disease	10	0	0
A. Mavrogordato, blood gases in certain diseases	10	0	0
D. Nabarro, continued investigations into the bacteriology of the human fæces	20	0	0
A. S. Grünbaum, further investigation of "preinfective" and "post-infective" substances in the blood	20	0	0
O. Rosenheim, (*a*) pancreatic lipase; (*b*) lipoids of the brain	20	0	0
M. C. Rosenheim, (*a*) chemical composition of spermine; (*b*) oxycholesterol in gall-stones	20	0	0
C. Bolton, pathology of dropsy of venous stagnation	10	0	0
J. Mellanby, enquiry into the nature of ferment actions	20	0	0
S. Vincent, certain aspects of "internal secretions"	25	0	0
E. P. Cumberbatch and J. W. Trevan, experimental investigation of the reaction of degeneration	36	0	0
D. F. Harris, nature and activities of tissue reductase	10	0	0
T. S. Hele, origin of ethereal sulphate	30	0	0
J. A. Gardner and G. W. Ellis, research on coprosterol and bihydrocholesterol and their derivatives	20	0	0
G. A. Buckmaster, enquiry into the nature of (*a*) methæmoglobin and its oxygen content; (*b*) relation of pure hæmochromogen to gases	15	0	0
Royal Society, for the purposes of the publications of the Society	500	0	0
Royal Society, towards assisting the scientific work of other societies	500	0	0
	£3,144	0	0

FROM THE COUNCIL FUND (Regulation 24).

	£	*s.*	*d.*
Royal Society, to defray expenses of experiments undertaken by War Committee	250	0	0

REGULATIONS GOVERNING THE USE OF THE LIBRARY OF THE ROYAL SOCIETY.

1. The Library shall be open to the Fellows every week-day (exclusive of Good Friday and Easter-eve, of Easter week, of a week at Whitsuntide, and of a week at Christmas), from 11 A.M. to 6 P.M., except on Saturdays, when it shall be open from 11 in the morning to 1 in the afternoon; but during the months of August and September, it shall be closed on week-days other than Saturdays at 4 P.M.

2. Any Fellow may have the loan of any of the printed Books of the Society, excepting such as the Council shall order not to be taken out of the Library; but he shall not be allowed to have in his possession more than ten volumes at a time. The loan of Manuscripts is exclusively vested in the President and Council.

3. A List of all Books and Manuscripts borrowed from the Library of the Royal Society, and of the Fellows of the Society to whom they are lent, shall be kept in the Library.

4. All books whatsoever belonging to the Society, shall be returned at a time to be specified by the Council in each year; and the Library shall be closed for one month after such time, or for such shorter periods as the Council may direct.

5. The value of such Books in the possession of any Fellow as are not returned to the Library, pursuant to the preceding Statute, shall be required to be paid by the person who has so detained them.

6. No persons other than Fellows have the privilege of using the Library, except upon a written introduction from a Fellow, with whom rests the responsibility for all books entrusted to the person introduced. Every such introduction shall be valid only until the 1st August next ensuing.

7. Dictionaries, Cyclopædias, and works of general reference do not circulate.

8. Books of exceptional rarity, size, or value, are only allowed to circulate by special permission of the Council.

9. All books are borrowed subject to recall after one month's interval.

10. All books are returnable to the Library *on the 1st August in each year*, and no books can be borrowed during the month of August.

11. All applications for the use of the Library are to be addressed to the *Assistant Secretary and Librarian*, who is charged with the carrying out of these regulations.

Ordered by the Library Committee at their meeting on the 16th December, 1898.

ADDITIONS TO LIBRARY, 1914–1915.

Adye (E. H.) Memoir on the Economic Geology of Navanagar State, in the Province of Kathiawar, India. 8vo. *Bombay* 1914. From the Author.

Advisory Committee on Aeronautics. Report on Gyroscopic Theory. By Sir G. Greenhill. Fol. *London* 1914. From the Committee.

Bayliss (W. M.) Principles of General Physiology. 8vo. *London* 1915. From the Author.

Bensaude (J.) L'Astronomie Nautique au Portugal à l'Epoque des Grandes Découvertes. 4to. *Berne* 1912. From the Author.

Brazil (V.) La Défense contre l'Ophidisme. 8vo. *Saint-Paul* 1914. From the Institut Sérumthérapique de S. Paulo.

British Antarctic Expedition, 1907–9 (Sir E. Shackleton). Reports on the Scientific Investigations. Geology. Vol. 1. 4to. *London* 1914. From the Publisher.

British Antarctic (*Terra Nova*) Expedition. Natural History Report. Zoology, Vol. 1, No. 1; Vol. 2, No. 1; Geology. Vol. 1, No. 1. 4to. *London* 1914. From the British Museum (Natural History).

Chappell (E.) Five-figure Mathematical Tables. 8vo. *London* 1915. From the Author.

Darboux (G.) Leçons sur la Théorie Générale des Surfaces. 2me Partie. 2me Edition. 8vo. *Paris* 1915. From the Author.

Deuxième Expédition Antarctique Française (1908–10), commandée par le Dr. J. Charcot. Sciences Physiques. 1 vol. Sciences Naturelles. 1 vol. 4to. *Paris*. From the Ministère de la Marine.

De Villamil (R.) Motion of Liquids. 8vo. *London* 1914. From the Author.

Galloway (W.) Great Colliery Explosions and their Means of Prevention. 8vo. *London* [1914.] From the Author.

Grassi (B.) Studi di un Zoologo sulla Malaria. 2a Edizione. 4to. *Roma* 1901. From H. G. Plimmer, F.R.S.

Histoire de la Science Nautique Portugaise à l'Epoque des Grandes Découvertes. Collection de Documents publiés par ordre du Ministère de l'Instruction Publique de la République Portugaise. Vols. 1, 2, 3, 5. 4to and fol. 1914–15. From the Portuguese Government.

Hull, (Edward, F.R.S.) Reminiscences of a Strenuous Life. 2nd Edition. 8vo. *London* 1910. From the Author.

Koningsberger (J. C.) Java, zoölogisch en biologisch. Pts. 1-10. 8vo. *Buitenzorg* 1911-14.
From the Department of Agriculture—Dutch East Indies.

Meyrick (E.) Exotic Microlepidoptera. Vol. 1, Pts. 1-15. 8vo. *London* 1912-15. From the Author.

Napier Tercentenary Celebration. Handbook of Exhibition of Napier Relics, &c. 8vo. *Edinburgh* 1914.
From the Royal Society, Edinburgh.

Rhijn (P. J. van) Derivation of Change of Colour with Distance and Apparent Magnitude. 4to. *Groningen* 1915. From the Author.

Scotia—Report on Work carried out by the S.S. *Scotia* 1913. Fol. *London* 1914. From the Meteorological Office.

Turner (Sir Wm.) Four volumes of Collected Papers. 4to. The Marine Mammals in the Anatomical Museum of the University of Edinburgh. 8vo. 1912. From the Author.

ADDITIONS TO SERIALS.

Journal of Agricultural Research. Vol. 1- , 1913- . 8vo. *Washington.* From the Department of Agriculture, Washington.

INSTITUTIONS ON WHICH THE ROYAL SOCIETY IS REPRESENTED.

UNIVERSITIES.

Oxford. Boards of Electors to the—	Representatives.	Appointed.
Savilian Professorship of Geometry	The President.	*Ex officio.*
Savilian Professorship of Astronomy.	"	"
Sedleian Professorship of Natural Philosophy.	"	"
Professorship of Experimental Philosophy.	"	"
Wykeham Professorship of Physics.	"	"
Waynflete Professorship of Chemistry.	"	"
Waynflete Professorship of Mineralogy.	"	"
Professorship of Geology.	"	"
Cambridge. Board of Electors to the—		
Lowndean Professorship of Astronomy and Geometry.	"	"
Liverpool University Court.	Prof. C. S. Sherrington.	Dec. 11, 1913.
Bristol University Court.	Sir Edward Fry	June 17, 1909.

PUBLIC SCHOOLS.

Governing bodies of:—		
Charterhouse School.	Sir William A. Tilden.	Oct. 29, 1908.
Christ's Hospital.	Professor H. E. Armstrong.	Jan. 16, 1896.
Dulwich College.	Dr. J. J. H. Teall.	Feb. 20, 1913.
Eton College.	Sir Henry Miers.	May 16, 1912.
Harrow School.	Sir Arch. Geikie.	June 16, 1892.
Rugby School.	Professor H. F. Newall.	July 6, 1911.
Shrewsbury School.	Sir Francis Darwin.	Oct. 22, 1908.
Westminster School.	Professor T. G. Bonney.	Oct. 27, 1881.
Winchester College.	Major P. A. MacMahon.	Feb. 16, 1899.

OTHER INSTITUTIONS.

Athenæum Club (Committee).	The President.	*Ex officio.*
British Museum (Trustee).	"	"
Chelsea Physick Garden.*	Prof. J. B. Farmer.	Feb. 21, 1907.
City and Guilds of London Institute (Governor).	The President.	*Ex officio.*
Hunterian Museum (Trustee).	"	"
Imperial Cancer Research Fund.	Sir John Bradford. Sir Watson Cheyne.	July 11, 1902. "
Imperial College of Science and Technology.	Mr. Horace Brown.	Mar. 23, 1911.
Lawes Agricultural Trust Managing Committee.	Prof. R. H. Biffen. Mr. Horace Brown. Prof. J. B. Farmer. Mr. Spencer Pickering.	
Lister Institute of Preventive Medicine.	Sir David Bruce.	Dec. 10, 1903.
Military Education (Advisory Board).	Sir William Abney.	April 30, 1903.
Plague in India (Advisory Committee).	Sir John Bradford. Prof. G. H. F. Nuttall.	Oct. 27, 1904. Feb. 28, 1907.
Sir John Soane's Museum.	Sir Clifford Allbutt.	Jan. 20, 1916.
Tropical Diseases Advisory Board.	Sir John Bradford.	Feb. 21, 1907.

The President of the Royal Society is also *ex officio* an honorary Member of the Royal Irish Academy.

* The History of the connection of the Physick Garden with the Royal Society will be found in the "Record of the Royal Society."

OTHER PUBLIC RESPONSIBILITIES OF THE ROYAL SOCIETY.

1. Government Grant for Scientific Investigations.—Administrators.

For the Regulations see page 110, *supra*. For the History of this Grant see "Record."

2. Kew Observatory.

A new scheme for the Management of Kew Observatory, with other Observatories, came into operation on July 1, 1910. Under this scheme the Gassiot Committee, appointed by the Royal Society to administer the Gassiot Trust Fund, is reconstituted as a Scientific Committee for the purpose of recommending observations and researches in magnetism, seismology, meteorology and geophysics to be carried out at Kew Observatory, Eskdalemuir and Valencia Observatories, which are now placed under the general direction of the Meteorological Committee (see below). The proceeds of the Gassiot Trust Fund, and the grant in aid made by the Treasury to the Royal Society for Eskdalemuir Observatory, will be used in such manner as the Gassiot Committee may approve for the maintenance of these observations and researches, in augmentation of the sums assigned for the purpose by the Meteorological Committee.

3. Meteorological Office.

The present Meteorological Committee is the official descendant of the Meteorological Department of the Board of Trade, the history of which is given in the Report by the Committee of Inquiry nominated by the Royal Society, the Board of Trade, and the Admiralty, respectively, which was printed and presented to Parliament in 1866. This Department was superseded in 1867 by the Meteorological Committee of the Royal Society. In 1877, the Committee transferred their charge to the Meteorological Council, a paid body, consisting of a chairman and four members, nominated by the President and Council of the Royal Society, and approved by the Lords Commissioners of the Treasury, with the Hydrographer of the Admiralty as an official member. In 1905 the Meteorological Council was replaced by a Committee appointed by H.M. Treasury, and containing two representatives of the Royal Society.

4. National Physical Laboratory.—Control vested in the President and Council of the Royal Society. (For scheme of organization see p. 103.)

5. Royal Observatory, Greenwich.—Visitors.

The Royal Society were appointed Visitors and Directors in 1710, a function which they continued to perform until the accession of King William IV, when, by the new warrant then issued, the President and six of the Fellows of the Royal Astronomical Society were added to the same number of the Royal Society to constitute the Board of Visitors, the President of the Royal Society being Chairman of the Board. A new warrant was issued by His Majesty King George V in 1911.

6. Standard Weights and Measures.—Custodians.

The Imperial Standard Yard and Pound in actual use for all important comparisons are at the Standards Office. Four copies of each of them are deposited in other places in case of injury or loss of the standards. One set of copies is in the custody of the Royal Society.

* LIST OF INSTITUTIONS
WHICH RECEIVE THE PHILOSOPHICAL TRANSACTIONS OR THE PROCEEDINGS OF THE ROYAL SOCIETY.

Institutions marked A are entitled to receive Philosophical Transactions, Series A, and Proceedings, Series A.

Institutions marked B are entitled to receive Philosophical Transactions, Series B, and Proceedings, Series B.

Institutions marked AB are entitled to receive Philosophical Transactions, Series A and B, and both series of Proceedings.

Institutions marked *p* are entitled to receive both series of Proceedings only, and those marked *pa*, *pb*, the A or B Series of Proceedings only.

Africa. (*See* EGYPT and UNION OF SOUTH AFRICA.)

America (North). (*See* MEXICO, UNITED STATES and CANADA.)

America (South).

Buenos Ayres.

AB. Museo Nacional.

Caracas.

B. University Library.

Cordova.

AB. Academia Nacional de Ciencias.

Demerara.

pb. Royal Agricultural and Commercial Society, British Guiana.

La Plata.

B. Museo de la Plata.

Rio de Janeiro.

pa. Observatorio.

Australia.

Adelaide.

p. Royal Society of South Australia.

AB. University.

Brisbane.

p. Royal Society of Queensland.

Hobart.

p. Royal Society of Tasmania.

Melbourne.

pa. Observatory.

p. Royal Society of Victoria.

AB. University Library.

Perth.

pa. Mines Department.

* A large number of these Exchanges are in suspense on account of the war.

Australia—*continued.*

Sydney.
- *p.* Australian Museum.
- *p.* Geological Survey.
- B. Linnean Society of New South Wales.
- AB. Royal Society of New South Wales.
- AB. University Library.

Austria.

Brünn.
- AB. Naturforschender Verein.

Cracow.
- AB. Kaiserliche Akademie der Wissenschaften.

Gratz.
- AB. Naturwissenschaftlicher Verein für Steiermark.

Innsbruck.
- AB. Das Ferdinandeum.
- *p.* Naturwissenschaftlich-Medicinischer Verein.

Prague.
- AB. Königliche Böhmische Gesellschaft der Wissenschaften.

Trieste.
- B. Museo di Storia Naturale.
- *p.* Società Adriatica di Scienze Naturali.

Vienna.
- *pb.* Anthropologische Gesellschaft.
- AB. Kaiserliche Akademie der Wissenschaften.
- *p.* K.K. Geographische Gesellschaft.
- AB. K.K. Geologische Reichsanstalt.
- B. K.K. Naturhistorisches Hof-Museum.
- B. K.K. Zoologisch-Botanische Gesellschaft.
- *pa.* Oesterreichische Gesellschaft für Meteorologie.
- A. Von Kuffner'sche Sternwarte.

Belgium.

Brussels.
- B. Académie Royale de Médecine.
- AB. Académie Royale des Sciences.
- B. Musée du Congo.
- B. Musée Royal d'Histoire Naturelle de Belgique.
- *pa.* Observatoire Royal.
- *p.* Société Belge de Géologie, de Paléontologie, et d'Hydrologie.
- *pb.* Société Malacologique de Belgique.

Ghent.
- AB. Université.

Belgium—*continued.*

Liége.

AB. Société des Sciences.

p. Société Géologique de Belgique.

Louvain.

B. Laboratoire de Botanique et de Biologie Cellulaire.

AB. Université.

Canada.

Edmonton.

AB. University of Alberta.

Fredericton, N.B.

p. University of New Brunswick.

Halifax, N.S.

p. Nova Scotian Institute of Science.

Hamilton.

p. Hamilton Scientific Association.

Kingston, Ontario.

AB. Queen's University.

Montreal.

AB. McGill University.

pb. Natural History Society.

Ottawa.

AB. Geological Survey of Canada.

AB. Royal Society of Canada.

AB. University.

St. John, N.B.

pb. Natural History Society.

Toronto.

pa. Royal Astronomical Society of Canada.

p. Canadian Institute.

AB. University.

Windsor, N.S.

p. King's College Library.

Ceylon.

Colombo.

B. Museum.

Denmark.

Copenhagen.

AB. Kongelige Danske Videnskabernes Selskab.

Egypt.

Alexandria.

AB. Bibliothèque Municipale.

Khartoum.

" Gordon College, Wellcome Laboratories.

England and Wales.

Aberystwyth.
AB. University College.

Bangor.
AB. University College of North Wales.

Birmingham.
AB. Central Free Library.
AB. University.
p. Philosophical Society.

Bolton.
p. Public Library.

Bristol.
p. Merchant Venturers' School.
AB. University College.

Cambridge.
AB. Philosophical Society.
p. Union Society.
AB. University Library.

Cardiff.
AB. Free Library.
AB. University College.

Chatham.
A. Royal Engineers' Head Quarters.

Dudley.
p. Dudley and Midland Geological and Scientific Society.

Englefield Green.
AB. Royal Holloway College.

Essex.
p. Essex Field Club.

Falmouth.
p. Royal Cornwall Polytechnic Society.

Greenwich.
A. Royal Observatory.

Harpenden.
AB. Lawes Agricultural Trust.

Kew.
B. Royal Gardens.

Leeds.
p. Philosophical Society.
AB. University.

Liverpool.
AB. Free Public Library.
p. Literary and Philosophical Society.
A. Observatory.
AB. University.
pb. School of Tropical Medicine.

England and Wales—*continued.*

London.

AB. Admiralty.
A. Board of Trade: Electrical Standards Laboratory.
pa. British Astronomical Association.
AB. British Museum.
AB. British Museum (Nat. Hist.).
AB. Chemical Society.
A. City and Guilds of London Institute.
pa. "Electrician," Editor of the.
B. Entomological Society.
AB. Geological Society.
AB. Geological Survey of Great Britain.
p. Geologists' Association.
AB. Guildhall Library.
A. Institution of Civil Engineers.
A. Institution of Electrical Engineers.
A. Institution of Mechanical Engineers.
A. Institution of Naval Architects.
pa. Iron and Steel Institute.
AB. King's College.
AB. King's Library.
B. Linnean Society.
p. Lister Institute of Preventive Medicine.
AB. London Institution.
p. London Library.
A. London Mathematical Society.
pa. Meteorological Office.
p. Pharmaceutical Society.
pa. Physical Society.
p. Quekett Microscopical Club.
p. Royal Agricultural Society.
p. Royal Anthropological Institute.
pb. Royal Army Medical College.
A. Royal Astronomical Society.
B. Royal College of Physicians.
B. Royal College of Surgeons.
p. Royal Engineers (for Libraries abroad, six copies).
p. Royal Geographical Society.
pb. Royal Horticultural Society.
p. Royal Institute of British Architects.
AB. Royal Institution of Great Britain.
A. Royal Meteorological Society.
p. Royal Microscopical Society.

England and Wales—*continued.*

AB. Royal Society of Arts.
B*p*. Royal Society of Medicine.
p. Royal Statistical Society.
AB. Royal United Service Institution.
A. "Science Abstracts."
p. Society of Chemical Industry (London Section).
pa. Standards Department, Board of Trade.
pb. Tropical Diseases Bureau.
AB. University College.
p. Victoria Institute.
AB. War Office.
p. War Office (Medical Department).
B. Zoological Society.

Manchester.
AB. Free Library.
AB. Literary and Philosophical Society.
p. Geological Society.
AB. University.

Newcastle.
AB. Armstrong College.
AB. Free Library.
pa. North of England Institute of Mining and Mechanical Engineers.
p. Society of Chemical Industry (Newcastle Section).

Norwich.
p. Norfolk and Norwich Literary Institution.

Nottingham.
AB. Central Free Public Library.

Oxford.
p. Ashmolean Society.
AB. Bodleian Library.
AB. Radcliffe Library.
A. Radcliffe Observatory.

Penzance.
p. Geological Society of Cornwall.

Plymouth.
B. Marine Biological Association.
p. Plymouth Institution.

Richmond.
A. National Physical Laboratory, Observatory Department.

Salford.
p. Royal Museum and Library.

Sheffield.
AB. University.

England and Wales—*continued.*

Southampton.

p. Hartley College.

Stonyhurst.

p. The College.

Swansea.

AB. Royal Institution.

Teddington.

A. National Physical Laboratory.

Woolwich.

AB. Royal Artillery Library.

Finland.

Helsingfors.

p. Societas pro Fauna et Flora Fennica.

AB. Société des Sciences.

France.

Bordeaux.

p. Académie des Sciences.

p. Faculté des Sciences.

pb. Société de Médecine et de Chirurgie.

p. Société des Sciences Physiques et Naturelles.

Caen.

pb. Société Linnéenne de Normandie.

Cherbourg.

p. Société des Sciences Naturelles.

Dijon.

p. Académie des Sciences.

Lille.

p. Faculté des Sciences.

Lyons.

AB. Académie des Sciences, Belles-Lettres et Arts.

AB. Université.

Marseilles.

AB. Faculté des Sciences.

Montpellier.

AB. Académie des Sciences et Lettres.

B. Faculté de Médecine.

Nantes.

p. Société des Sciences Naturelles de l'Ouest de la France.

Paris.

AB. Académie des Sciences de l'Institut.

pb. Année Biologique (M. H. DE VARIGNY).

p. Association Française pour l'Avancement des Sciences.

pa. Bureau des Longitudes.

France—*continued.*

A. Bureau International des Poids et Mesures.
pa. Commission des Annales des Ponts et Chaussées.
pa. Conservatoire des Arts et Métiers.
p. Cosmos (M. L'ABBÉ VALETTE).
AB. École des Mines.
AB. École Normale Supérieure.
AB. École Polytechnique.
AB. Faculté des Sciences à la Sorbonne.
B. Institut Pasteur.
B. Jardin des Plantes.
pa. "Le Radium."
A. L'Observatoire.
pa. Observatoire d'Astronomie Physique de Meudon.
p. Revue Générale des Sciences.
p. Revue Scientifique.
AB. Service Hydrographique de la Marine.
pb. Société de Biologie.
AB. Société d'Encouragement pour l'Industrie Nationale.
AB. Société de Géographie.
pa. Société de Physique.
AB. Société Géologique.
pa. Société Mathématique.

Rennes.
p. Université.

Toulouse.
AB. Académie des Sciences.
A. Faculté des Sciences.

Germany.

Berlin.
A. Association Géodésique International.
A. Deutsche Chemische Gesellschaft.
pa. Gesellschaft für Erdkunde.
pb. Gesellschaft Naturforschender Freunde.
pa. Königliches Materialprüfungsamt.
AB. Königliche Preussische Akademie der Wissenschaften.
A. Königliche Sternwarte.
A. Deutsche Physikalische Gesellschaft.

Bonn.
AB. Universität.

Bremen.
p. Naturwissenschaftlicher Verein.

Breslau.
p. Schlesische Gesellschaft für Vaterländische Kultur.

Germany—*continued.*

Brunswick.
p. Verein für Naturwissenschaft.

Carlsruhe. See Karlsruhe.

Charlottenburg.
A. Physikalisch-Technische Reichsanstalt.

Danzig.
AB. Naturforschende Gesellschaft.

Dresden.
p. Verein für Erdkunde.

Emden.
p. Naturforschende Gesellschaft.

Erlangen.
AB. Physikalisch-Medicinische Societät.

Frankfurt-am-Main.
AB. Senckenbergische Naturforschende Gesellschaft.
pb. Zoologische Gesellschaft.

Frankfurt-am-Oder.
p. Naturwissenschaftlicher Verein.

Freiburg-im-Breisgau.
AB. Universität.

Giessen.
AB. Grossherzogliche Universität.

Görlitz.
p. Naturforschende Gesellschaft.

Göttingen.
AB. Königliche Gesellschaft der Wissenschaften.

Halle.
AB. Kaiserliche Leopoldinisch-Carolinische Deutsche Akademie der Naturforscher.
pb. Naturwissenschaftlicher Verein für Sachsen und Thüringen.

Hamburg.
pb. Naturhistorisches Museum.
AB. Naturwissenschaftlicher Verein.

Heidelberg.
A. Grossherzogliche Sternwarte, Astronomisches Institut.
AB. Universität.

Jena.
AB. Medicinisch-Naturwissenschaftliche Gesellschaft.

Karlsruhe.
p. Technische Hochschule.

Kiel.
A. Astronomische Nachrichten.
p. Naturwissenschaftlicher Verein für Schleswig-Holstein.

Germany—*continued.*

pa. Sternwarte.
AB. Universität.

Königsberg.
AB. Königliche Physikalisch-Ökonomische Gesellschaft.

Leipsic.
A. "Annalen der Physik."
pa. "Jahrbuch der Drahtlosen Telegraphie und Telephonie."
AB. Königlich Sächsische Gesellschaft der Wissenschaften.

Magdeburg.
p. Naturwissenschaftlicher Verein.

Marburg.
AB. Universität.

Munich.
AB. Königlich Bayerische Akademie der Wissenschaften.
pb. Zeitschrift für Biologie.

Münster.
AB. Königliche Universität.

Potsdam.
A. Astrophysikalisches Observatorium.

Rostock.
AB. Universität.

Strasburg.
AB. Universität.

Tübingen.
AB. Universität.

Würzburg.
AB. Physikalisch-Medicinische Gesellschaft.

Greece.

Athens.
A. National Observatory.

Holland. (*See* NETHERLANDS.)

Hungary.

Buda-pest.
AB. Á Magyar Tudós Társaság. Die Ungarische Akademie der Wissenschaften.
p. Königl. Ungarische Geologische Anstalt.

Hermannstadt.
p. Siebenbürgischer Verein für die Naturwissenschaften.

Klausenburg.
AB. Az Erdélyi Muzeum. (Das Siebenbürgische Museum.)

Schemnitz.
p. K. Ungarische Berg- und Forst-Akademie.

India.

Bombay.

AB. Elphinstone College.

pa. Government Observatory.

p. Royal Asiatic Society (Bombay Branch).

Calcutta.

AB. Asiatic Society of Bengal.

AB. Geological Museum.

pa. Great Trigonometrical Survey of India.

AB. Indian Museum.

pa. Mathematical Society.

pa. The Director General of Observatories.

Madras.

B. Central Museum.

A. Kodaikanal Observatory.

Roorkee.

p. Roorkee College.

Simla.

pa. India Meteorological Department.

Ireland.

Armagh.

A. Observatory.

Belfast.

AB. Queen's University.

Cork.

AB. University College.

Dublin.

AB. National Library of Ireland.

A. Observatory.

B. Royal College of Surgeons in Ireland.

AB. Royal Dublin Society.

AB. Royal Irish Academy.

AB. Trinity College.

p. University College.

Galway.

AB. University College.

Italy.

Acireale.

p. Accademia di Scienze, Lettere ed Arti.

Bologna.

AB. Accademia delle Scienze dell' Istituto.

Italy—*continued.*

Catania.

AB. Accademia Gioenia di Scienze Naturali.

Florence.

p. Biblioteca Nazionale Centrale.

AB. Museo Botanico.

p. Reale Istituto di Studi Superiori.

Genoa.

p. R. Università.

p. Società Ligustica di Scienze Naturali e Geografiche.

Milan.

AB. Reale Istituto Lombardo di Scienze, Lettere ed Arti.

AB. Società Italiana di Scienze Naturali.

Monte Rosa.

pb. Laboratoire International de Physiologie.

Naples.

pb. Istituto Zoologico, R. Università.

p. Società di Naturalisti.

AB. Società Reale, Accademia delle Scienze.

B. Stazione Zoologica.

Padua.

p. University.

Palermo.

A. Circolo Matematico.

AB. Consiglio di Perfezionamento (Società di Scienze Naturali ed Economiche).

A. Reale Osservatorio.

Pisa.

pa. Il Nuovo Cimento.

p. Società Toscana di Scienze Naturali.

Rome.

p. Accademia Pontificia de' Nuovi Lincei.

pb. Institut International d'Agriculture.

AB. Reale Accademia dei Lincei.

A. Reale Ufficio Centrale di Meteorologia e di Geodinamica. Collegio Romano.

p. R. Comitato Geologico d' Italia.

AB. Società Italiana delle Scienze.

A. Specola Vaticana.

Sassari.

p. Università. Istituto Fisiologico.

Siena.

p. Reale Accademia dei Fisiocritici.

Italy—*continued.*

Turin.

AB. Biblioteca Nazionale.

pb. Laboratorio di Fisiologia.

AB. Reale Accademia delle Scienze.

Venice.

p. Ateneo Veneto.

AB. Reale Istituto Veneto di Scienze, Lettere ed Arti.

Japan.

Tokiô.

p. Asiatic Society of Japan.

AB. Imperial University.

Java.

Batavia.

p. K. Natuurkundige Vereeniging, Weltevreden.

Buitenzorg.

pb. Jardin Botanique.

Luxembourg.

Luxembourg.

p. Institut Grand-Ducal.—Section des Sciences Naturelles.

Malta.

p. Public Library.

Mauritius.

A. Royal Alfred Observatory.

p. Royal Society of Arts and Sciences.

Mexico.

p. Sociedad Científica "Antonio Alzate."

Netherlands.

Amsterdam.

AB. Koninklijke Akademie van Wetenschappen.

pb. K. Zoologisch Genootschap "Natura Artis Magistra."

Haarlem.

AB. Hollandsche Maatschappij der Wetenschappen.

p. Musée Teyler.

Leyden.

AB. University.

Rotterdam.

AB. Bataafsch Genootschap der Proefondervindelijke Wijsbegeerte.

Utrecht.

AB. Provinciaal Genootschap van Kunsten en Wetenschappen.

New Zealand.

Wellington.

AB. New Zealand Geological Survey.

AB. New Zealand Institute.

Norway.

Bergen.

AB. Bergenske Museum.

Christiania.

AB. Kongelige Norske Frederiks Universitet.

Tromsoe.

p. Museum.

Trondhjem.

AB. Kongelige Norske Videnskabers Selskab.

Philippine Islands.

Manila.

pb. Bureau of Science.

Portugal.

Coimbra.

AB. Universidade.

Lisbon.

AB. Academia de Sciencias di Lisboa.

pb. R. Instituto Bacteriologico, Camara Pestana.

p. Secção dos Trabalhos Geologicos de Portugal.

Oporto.

p. Annaes de Sciencias Naturaes.

Russia.

Dorpat (Juriew).

AB. Université.

Ekaterinoslav.

pa. School of Mines.

Irkutsk.

pa. Observatoire.

p. Société Impériale Russe de Géographie (Section de la Sibérie Orientale).

Kazan.

AB. Imperatorsky Kazansky Universitet.

pa. Société Physico-Mathématique.

Kharkoff.

pb. Section Médicale de la Société des Sciences Expérimentales, Université de Kharkow.

Kieff.

p. Société des Naturalistes.

Russia—*continued.*

Kronstadt.

pa. Compass Observatory.

Moscow.

AB. Le Musée Public.

B. Société Impériale des Naturalistes.

Odessa.

p. Société des Naturalistes de la Nouvelle-Russie.

Petrograd.

AB. Académie Impériale des Sciences.

B. Archives des Sciences Biologiques.

AB. Comité Géologique.

A. Commission Sismique Centrale.

AB. Ministère de la Marine.

A. Observatoire Physique Central.

Pulkowa.

A. Nikolai Haupt-Sternwarte.

Scotland.

Aberdeen.

AB. University.

Edinburgh.

AB. Advocates' Library.

p. Geological Society.

p. Royal College of Physicians (Research Laboratory).

pb. Royal Medical Society.

A. Royal Observatory.

p. Royal Physical Society.

p. Royal Scottish Society of Arts.

AB. Royal Society.

Glasgow.

AB. Mitchell Free Library.

pb. Natural History Society.

p. Royal Philosophical Society.

AB. University.

Langholm.

A. Eskdalemuir Observatory.

Servia.

Belgrade.

p. Académie Royale de Serbie.

Spain.

Cadiz.

A. Instituto y Observatorio de Marina de San Fernando.

Spain—*continued.*

Madrid.

p. Comisión del Mapa Geológico de Espãna.

AB. Real Academia de Ciencias.

Sweden.

Gottenburg.

AB. Kongl. Vetenskaps och Vitterhets Samhälle.

Lund.

AB. Universitet.

Stockholm.

A. Acta Mathematica.

AB. Kongliga Svenska Vetenskaps-Akademie.

AB. Sveriges Geologiska Undersökning.

Upsala.

AB. Uuiversitet.

Switzerland.

Basel.

p. Naturforschende Gesellschaft.

Bern.

p. Naturforschende Gesellschaft.

AB. Société Hélvétique des Sciences Naturelles.

Geneva.

AB. Institut National Genevois.

pa. Journal de Chimie Physique.

AB. Société de Physique et d'Histoire Naturelle.

Lausanne.

p. Société Vaudoise des Sciences Naturelles.

Neuchâtel.

p. Société des Sciences Naturelles.

Zürich.

AB. Das Schweizerische Polytechnikum.

p. Naturforschende Gesellschaft.

pa. Sternwarte.

Union of South Africa.

Bloemfontein.

p. Grey University College.

Cape Town.

A. Royal Observatory.

AB. South African Library.

B. South African Museum.

p. Royal Society of South Africa.

Union of South Africa—*continued.*

Grahamstown.

p. Public Library.

Johannesburg.

A. Union Observatory.

United States.

Albany.

AB. New York State Library.

Annapolis.

AB Naval Academy.

Austin (Texas).

p. Texas Academy of Sciences.

Baltimore.

AB. Johns Hopkins University.

Berkeley.

p. University of California.

Boston.

AB. American Academy of Sciences.

B. Boston Society of Natural History.

A. Massachusetts Institute of Technology.

Brooklyn.

AB. Brooklyn Library.

Cambridge (Mass.).

AB. Harvard University.

B. Museum of Comparative Zoology.

Chapel Hill (N.C.).

p. Elisha Mitchell Scientific Society.

Charleston.

p. Charleston Museum.

Chicago.

AB. Academy of Sciences.

pa. Astrophysical Journal.

pb. Field Museum of Natural History.

A. Yerkes Observatory (University of Chicago).

Davenport (Iowa).

p. Academy of Natural Sciences.

Ithaca (N.Y.).

A. Journal of Physical Chemistry.

pa. Physical Review (Cornell University).

Lawrence.

p. Kansas University.

United States—*continued.*

Madison.

p. Wisconsin Academy of Sciences.

Mount Hamilton (California).

A. Lick Observatory.

New Haven (Conn.).

AB. American Journal of Science.

AB. Connecticut Academy of Arts and Sciences.

New York.

p. American Geographical Society.

A. American Mathematical Society.

pb. American Museum of Natural History.

AB. Columbia University Library.

AB. New York Academy of Sciences.

pb. New York Medical Journal.

Pasadena.

A. Carnegie Institution Solar Observatory.

Philadelphia.

AB. Academy of Natural Sciences.

AB. American Philosophical Society.

p. Franklin Institute.

p. University of Pennsylvania.

Rochester (N.Y.).

p. Academy of Science.

St. Louis.

p. Academy of Science.

San Francisco.

AB. California Academy of Sciences.

Washington.

p. American Association for the Advancement of Science.

A. Bureau of Standards.

AB. Patent Office.

AB. Smithsonian Institution.

A. United States Coast Survey.

B. United States Bureau of Fisheries.

p. United States Department of Agriculture.

A. United States Department of Agriculture (Weather Bureau).

AB. United States Geological Survey.

AB. United States Naval Observatory.

West Point (N.Y.).

AB. United States Military Academy.

THE CROONIAN LECTURES.

LIST OF LECTURERS AND SUBJECTS.

1738. Alexander Stuart.

"On the Motion of the Heart, founded on some Anatomical Observations and Experiments."
Phil. Trans., Vol. 40, Supplement; Vol. 41, p. 675.

1739. Frank Nicholls.

"An Enquiry into Muscular Motion."
Journal Book, Vol. XVIII. p. 70.

1740. Alexander Stuart.

"On the Peristaltic Motion of the Intestines."
Journal Book, Vol. XVIII. p. 227—9.

1740. Alexander Stuart.

"Microscopical Observations on several parts of live Frogs."
Journal Book, Vol. XVIII. p. 290.

1741. James Douglas.

"Description of the several Muscles, Membranes and parts belonging to the Uvula of the Palate, and concerned in its action; as also of the several parts subservient to the uses of the Tuba Eustachiana."
Journal Book, Vol. XVIII. p. 377.

1742. James Douglas.

"Description and Structure of the Human Bladder, with the Uses of its Muscles and Membranes."
Journal Book, Vol. XVIII. p. 419.

1744.* James Parsons.

"An Introductory Discourse on Muscular Motion."
Phil. Trans., Vol. 43, Supplement.

1745. James Parsons.

"On Muscular Motion." Phil. Trans., Vol. 43, Supplement.

1746. James Parsons.

"Description of the several Muscles of the Face; with their particular Functions and Uses."
Phil. Trans., Vol. 44, Part I., Supplement.

1747. Browne Langrish.

"On the Theory of Muscular Motion."
Phil. Trans., Vol. 44, Part II., Supplement.

* Lecture revived, "the deficiency of the Fund being made good by the Rents."

1750. James Parsons.

"On Muscular Motion." Journal Book, Vol. XXI. p. 357.

1751. James Parsons.

"Critical Remarks upon the Motion and Uses of the Human Pelvis." Journal Book, Vol. XXI. pp. 641, 652.

1752 and 1753. Not recorded.

1754 to 1758. Charles Morton.

Minutes of Council, Vol. IV. pp. 168, 189: and Annual Accounts.

1759 and 1760. Not recorded.

1761. Charles Morton. (?)

Annual Accounts.

1762 to 1774. Not recorded.

1775 and 1776. John Hunter.

Minutes of Council, Vol. VI. pp. 283, 335.

1777 to 1781. John Hunter.

"On the Construction and Application of Muscles and the Power by which they are actuated."
Journal Book, Vol. XXXI. p. 194: and Annual Accounts, 1781.

1782. John Hunter.

"On the Density and Firmness of a Muscle as contributing to its Strength and Agility."
Journal Book, Vol. XXXI. p. 194.

1783. Not recorded.

1784. Foart Simmons.

"On the Irritability of the Muscular Fibres."
Journal Book, Vol. XXXI. p. 642.

1785. Edward Whittaker Grey.

"An Examination into Haller's Theory of Muscular Motion."
Journal Book, Vol. XXXII. p. 259.

1786. Edward Whittaker Grey.

"On the Effects of different kinds of Salts applied as Stimulants on the Muscles." Journal Book, Vol. XXXII. p. 468.

1787. George Fordyce.

"On Muscular Motion." Phil. Trans., Vol. 78, p. 23.

1788. Sir Gilbert Blane, Bart.

"On the Nature of the Muscles, and on the Theory of Muscular Motion." Journal Book, Vol. XXXIII. p. 268.

1789. Sir William Blizard.

"On the Theory of Muscular Motion."
Journal Book, Vol. XXXIV. p. 9.

1790. Sir Everard Home, Bart.

"On the Mechanism employed in producing Muscular Motion."
Journal Book, Vol. XXXIV. p. 200.

1791. Matthew Baillie.

"A general view of the Nature of the Muscles, and an enumeration of the most striking facts connected with the Theory of their Motion." Journal Book, Vol. XXXIV. p. 419.

1792. Not recorded.

1793. Sir Everard Home, Bart.*

"On Mr. Hunter's Experiments to ascertain whether the Crystalline Humour of the Eye be muscular."
Journal Book, Vol. XXXV. p. 166.

1794. Sir Everard Home, Bart.

"On the Crystalline Humour of the Eye." Phil. Trans. 1795

1795. Sir Everard Home, Bart.

"On the Mechanism employed in producing Muscular Motion."
Phil. Trans. 1795.

1796. Sir Everard Home, Bart.

"On the Crystalline Humour of the Eye." Phil. Trans. 1796.

1797. John Abernethy.

"A general Review of the latest Opinions relative to Animal Life and Motion." Journal Book, Vol. XXXVI. p. 340.

1798. Sir Everard Home, Bart.

"Experiments and Observations upon the Structure of Nerves."
Phil. Trans. 1799.

1799. Sir Everard Home, Bart.

"On the Structure and Uses of the Membrana Tympani."
Phil. Trans. 1800.

1800. Sir Everard Home, Bart.

"On the Irritability of Nerves." Phil. Trans. 1801.

1801. Sir Everard Home, Bart.

"On the power of the Eye to adjust itself to different distances when deprived of the Crystalline Lens."
Phil. Trans. 1802.

* The decease of Mr. Hunter took place before the Lecture, on which he was engaged by appointment of the Council, was completed.

1802. Not recorded.

1803. John Pearson.

"On Muscular Motion."

Journal Book, Vol. XXXVIII. p. 137.

1804. Sir Anthony Carlisle.

"On Muscular Motion." Phil. Trans. 1805.

1805. Sir Anthony Carlisle.

"On the Arrangement and Mechanical Action of the Muscles of Fishes." Phil. Trans. 1806.

1806. John Pearson.

"Remarks on Muscular Power, and on some of the circumstances by which it is increased, diminished, or finally abolished."

Journal Book, Vol. XXXIX. p. 176.

1807. Sir Anthony Carlisle.

"On the Natural History and Chemical Analysis of the substances which constitute the Muscles of Animals."

Journal Book, Vol. XXXIX. p. 451.

1808 Thomas Young.

"On the Functions of the Heart and Arteries."

Phil. Trans. 1809.

1809. William Hyde Wollaston.

"Observations on the Mode of Action of Voluntary Muscles, and on the causes which derange, and assist, the Action of the Heart and Blood Vessels." Phil. Trans. 1810.

1810. Benjamin Collins Brodie.

"Physiological Researches, respecting the Influence of the Brain on the Action of the Heart, and on the Generation of Animal Heat." Phil. Trans. 1811.

1811 and 1812. Not recorded.

1813. Benjamin Collins Brodie.

"On the Influence of the Nervous System on the Action of the Muscles in general and of the Heart in particular."

Journal Book, Vol. XLI. p. 347.

1814 to 1816. Not recorded.

1817. Sir Everard Home, Bart.

"On the Changes the Blood undergoes in the act of Coagulation."

Phil. Trans. 1818.

1818. Sir Everard Home, Bart.

"On the Conversion of Pus into Granulations, or New Flesh."

Phil. Trans. 1819.

1819. Sir Everard Home, Bart.

"A further Investigation of the component parts of the Blood."

Phil. Trans. 1820.

1820. Sir Everard Home, Bart.

"Microscopical Observations on the following subjects:—On the Brain and Nerves; showing that the Materials of which they are composed exist in the Blood; on the Discovery of Valves in the branches of the *vas breve*, lying between the villous and muscular coats of the Stomach; on the Structure of the Spleen." Phil. Trans. 1821.

1821. Sir Everard Home, Bart.

"On the Anatomical Structure of the Eye; illustrated by Microscopical Drawings, executed by F. Bauer." Phil. Trans. 1822.

1822. Francis Bauer.

"Microscopical Observations on the Suspension of the Muscular Motions of the Vibrio Tritici." Phil. Trans. 1823.

1823. Sir Everard Home, Bart.

"On the Internal Structure of the Human Brain, when examined in the Microscope, as compared with that of Fishes, Insects and Worms." Phil. Trans. 1824.

1824. Sir Everard Home, Bart.

"On the existence of Nerves in the Placenta." Phil. Trans. 1825.

1825. Sir Everard Home, Bart.

"On the Structure of a Muscular Fibre from which are derived its Elongation and Contraction." Phil. Trans. 1826.

1826. Sir Everard Home, Bart.

"An Enquiry into the mode by which the Propagation of the Species is carried on, in the Common Oyster, and in the large Fresh-water Muscle." Phil. Trans. 1827.

1827. Sir Everard Home, Bart.

"On the Muscles peculiar to Organs of Sense in particular Quadrupeds and Fishes." Journal Book, Vol. XLV. p. 143.

1828. Not appointed.

1829. Sir Everard Home, Bart.

"A Report on the Peculiarities met with in the Stomach of the Zariffa." Journal Book, Vol. XLV. p. 580.

1830 to 1850. Not appointed.

1851. Richard Owen.

"On the Megatherium." Proceedings, Vol. 6.

1852–1856. Not appointed.

1857. James Paget.

"On the Cause of the Rhythmic Action of the Heart." Proceedings, Vol. 8.

1858. Thomas Henry Huxley.

"On the Theory of the Vertebrate Skull." Proceedings, Vol. 9.

1859. Not appointed.

1860. James Bell Pettigrew.

"On the Arrangement of the Muscular Fibres of the Ventricular Portion of the Heart of the Mammal."
Proceedings, Vol. 10.
Phil. Trans. 1864.

1861. Charles Édouard Brown-Séquard.

"On the Relations between Muscular Irritability, Cadaveric Rigidity, and Putrefaction." Proceedings, Vol. 11.

1862. Albert Kölliker.

"On the Termination of Nerves in Muscles, as observed in the Frog: and on the disposition of the Nerves in the Frog's Heart." Proceedings, Vol. 12.

1863. Joseph Lister.

"On the Coagulation of the Blood." Proceedings, Vol. 12.

1864. Hermann Helmholtz.

"On the Normal Motions of the Human Eye in relation to Binocular Vision." Proceedings, Vol. 13.

1865. Lionel S. Beale.

"On the ultimate Nerve-fibres distributed to Muscle and some other Tissues, with Observations upon the Structure and probable Mode of Action of a Nervous Mechanism."
Proceedings, Vol. 14.

1866. Not appointed.

1867. J. S. Burdon-Sanderson.

"On the Influence exercised by the Movements of Respiration on the Circulation of the Blood." Phil. Trans. Vol. 157.

1868. Not appointed.

1869. Not appointed.

1870. Augustus V. Waller.

"On the Results of the Method, introduced by the Author, of investigating the Nervous System, more especially as applied to the Elucidation of the Functions of the Pneumogastric and Sympathetic Nerves in Man."
Proceedings, Vol. 18.

1871 and 1872. Not appointed.

1873. Benjamin Ward Richardson.

"On Muscular Irritability after Systemic Death."
Proceedings, Vol. 21.

1874. David Ferrier.
"The Localization of Function in the Brain."
Proceedings, Vol. 22 (Abstr.).

1875. David Ferrier.
"Experiments on the Brain of Monkeys. Second Series."
Phil. Trans. 1875.

1876. G. J. Romanes.
"Preliminary Observations on the Locomotor System of Medusæ." Phil. Trans. 1876.

1877 J. S. Burdon-Sanderson and F. J. M. Page.
"On the Mechanical Effects, and on the Electrical Disturbance, consequent on Excitation of the Leaf of *Dionæa muscipula.*" Proceedings, Vol. 25.

1878. H. N. Moseley.
"On the Structure of the Stylasteridæ: a Family of the Hydroid Stony Corals." Phil. Trans. 1878.

1879. W. K. Parker.
"On the Structure and Development of the Skull in the Lacertilia. Part I. On the Skull of the Common Lizards (*Lacerta agilis*, *L. viridis*, and *Zootoca vivipara*)."
Phil. Trans. 1879.

1880. Rev. S. Haughton.
"On some Elementary Principles in Animal Mechanics, No. IX. The Relation between the Maximum Work done, the Time of Lifting, and the Weights lifted by the Arms."
Proceedings, Vol. 30.

1881. G. J. Romanes and J. C. Ewart.
"Observations on the Locomotor System of Medusæ."
Phil. Trans. 1881.

1882. W. H. Gaskell.
"On the Rhythm of the Heart of the Frog, and on the Nature of the Action of the Vagus Nerve."
Phil. Trans. 1882.

1883. H. N. Martin.
"On the Direct Influence of Gradual Variations of Temperature upon the Rate of Beat of the Dog's Heart."
Phil. Trans. 1883.

1884 and 1885. Not appointed.

1886. L. C. Wooldridge.
"The Coagulation of the Blood." Proceedings, Vol. 40.

1887. H. G. Seeley.
"On *Pareiasaurus bombidens* (Owen) and the Significance of its Affinities to Amphibians, Reptiles, and Mammals."
Phil. Trans., B, 1888

1888. W. Kühne (Heidelberg).
"Ueber die Entstehung der Vitalen Bewegung."
Proceedings, Vol. 44.

1889. Dr. Roux (Institut Pasteur).
"Les Inoculations Préventives." Proceedings, Vol. 46.

1890. H. Marshall Ward.
"The Relations between Host and Parasite in certain Epidemic Diseases of Plants." Proceedings, Vol. 47.

1891. Francis Gotch and Victor Horsley.
"On the Mammalian Nervous System; its Functions and their Localization determined by an Electrical Method."
Phil. Trans., B, Vol. 182.

1892. Angelo Mosso (Turin).
"Les Phénomènes psychiques et la Température du Cerveau."
Phil. Trans., B, Vol. 183.

1893. Rudolf Virchow (Berlin).
"The Position of Pathology among Biological Studies."
Proceedings, Vol. 53.

1894. S. Ramón y Cajal (Madrid).
"La fine Structure des Centres Nerveux." Proceedings, Vol. 55.

1895. T. W. Engelmann (Utrecht).
"On the Nature of Muscular Contraction."
Proceedings, Vol. 57.

1896. Augustus D. Waller.
"Observations on Isolated Nerve." Phil. Trans., B, Vol. 188.

1897. Charles S. Sherrington.
"The Mammalian Spinal Cord as an Organ of Reflex Action."
Phil. Trans., B, Vol. 190.

1898. Wilhelm Pfeffer (Leipzig).
"The Nature and Significance of Functional Metabolism in the Plant." Proceedings, Vol. 63.

1899. J. S. Burdon-Sanderson.
"On the Relation of Motion in Animals and Plants to the Electrical Phenomena which are associated with it."
Proceedings, Vol. 65.

1900. Paul Ehrlich (Frankfort-on-M.).
"On Immunity with Special Reference to Cell Life."
Proceedings, Vol. 66.

1901. C. Lloyd Morgan.
"Studies in Visual Sensation." Proceedings, Vol. 68.

1902. Arthur Gamgee.
"On certain Chemical and Physical Properties of Hæmoglobin."
Proceedings, Vol. 70.

1903. C. Timiriazeff.
"The Cosmical Function of the Green Plant."
Proceedings, Vol. 72.

1904. Ernest Henry Starling and William Maddock Bayliss.
"The Chemical Regulation of the Secretory Process."
Proceedings, Vol. 73.

1905. William Bate Hardy.
"On the Globulins." Proceedings, B, Vol. 79.

1906. John Newport Langley.
"On Nerve Endings and on Special Excitable Substances in Cells."
Proceedings, B, Vol. 78.

1907. John Bretland Farmer.
"Structural Constituents of the Nucleus, and their Relation to the Organization of the Individual."
Proceedings, B, Vol. 79.

1908. Gustaf Retzius (Stockholm).
"The Principles of the Minute Structure of the Nervous System as revealed by Recent Investigations."
Proceedings, B, Vol. 80.

1909. Edward Albert Schäfer.
"The Functions of the Pituitary Body."
Proceedings, B, Vol. 81.

1910. G. Klebs.
"Alterations in the Development and Forms of Plants as a Result of Environment." Proceedings, B, Vol. 82.

1911. Thomas Gregor Brodie.
"A New Conception of the Glomerular Function."
Proceedings, B, Vol. 87.

1912. Keith Lucas.
"The Process of Excitation in Nerve and Muscle."
Proceedings, B, Vol. 85.

1913. Robert Broom.
"The Origin of Mammals."
Proceedings, B, Vol. 87.

1914. Edmund B. Wilson.
"The Bearing of Cytological Research on Heredity."
Proceedings, B, Vol. 88.

1915. Walter Morley Fletcher and Frederick Gowland Hopkins.
"The Respiratory Process in Muscle; and the Nature of Muscular Motion."

THE BAKERIAN LECTURES.

List of Lecturers and Subjects.

1775. Peter Woulfe.

"Experiments made in order to ascertain the nature of some Mineral Substances, and in particular to see how far the Acids of Sea-Salt and of Vitriol contribute to Mineralize Metallic and other Substances."—Part I.
Journal Book, Vol. XXIX. p. 135.

1776 and 1777. Peter Woulfe.

1778. John Ingen-Housz.

"Electrical Experiments to explain how far the Phenomena of the Electrophorus may be accounted for by Dr. Franklin's Theory of Positive and Negative Electricity."
Phil. Trans., Vol. 68.

1779. John Ingen-Housz.

"Improvements in Electricity." Phil. Trans., Vol. 69.

1780. Tiberius Cavallo.

"Thermometrical Experiments and Observations."
Phil. Trans., Vol. 70.

1781. Tiberius Cavallo.

"An Account of some Thermometrical Experiments."
Phil. Trans., Vol. 71.

1782. Tiberius Cavallo.

"An Account of some Experiments relating to the Property of Common and Inflammable Airs of pervading the Pores of Paper." Journal Book, Vol. XXXI. p. 203.

1783. Tiberius Cavallo.

"Description of an improved Air Pump."
Journal Book, Vol. XXXI. p. 401.

1784. Tiberius Cavallo.

"An Account of some Experiments made with the new improved Air Pump.' Journal Book, Vol. XXXI. p. 631.

1785. Tiberius Cavallo.

"Magnetical Experiments and Observations."
Phil. Trans., Vol. 76.

1786. Tiberius Cavallo.

"Magnetical Experiments and Observations."
Phil. Trans., Vol. 77.

1787. Tiberius Cavallo.

"Of the Methods of manifesting the Presence, and ascertaining the Quality, of small Quantities of Natural or Artificial Electricity." Phil. Trans., Vol. 78.

1788. Tiberius Cavallo.

"On an Improvement in the Blow Pipe."
Journal Book, Vol. XXXIII. p. 257.

1789. Tiberius Cavallo.

"Magnetical Experiments and Observations."
Journal Book, Vol. XXXIV. p. 3.

1790. Tiberius Cavallo.

"A Description of a new Pyrometer."
Journal Book, Vol. XXXIV. p. 208.

1791. Tiberius Cavallo.

"On the Method of Measuring Distances by means of Telescopes furnished with Micrometers."
Journal Book, Vol. XXXIV. p. 429.

1792. Tiberius Cavallo.

"An Account of the Discoveries concerning Muscular Motion, which have been lately made, and are commonly known by the name of Animal Electricity."
Journal Book, Vol. XXXIV. p. 609.

1793. George Fordyce.

"An Account of a New Pendulum." Phil. Trans. 1794.

1794. Samuel Vince.

"Observations on the Theory of the Motion and Resistance of Fluids; with a Description of the Construction of Experiments, in order to obtain some Fundamental Principles."
Phil. Trans. 1795.

1795 and 1796. Samuel Vince. (?)

1797. Samuel Vince.

"Experiments upon the Resistance of Bodies moving in Fluids."
Phil. Trans. 1798.

1798. Samuel Vince.

"Observations upon an unusual Horizontal Refraction of the Air; with Remarks on the Variations to which the lower Parts of the Atmosphere are sometimes subject."
Phil. Trans. 1799.

1799. Samuel Vince. (?)

1800. Thomas Young.

"On the Mechanism of the Eye." Phil. Trans. 1801.

1801. Thomas Young.

"On the Theory of Light and Colours." Phil. Trans. 1802.

1802. William Hyde Wollaston.

"Observations on the Quantity of Horizontal Refraction; with Method of measuring the Dip at Sea." Phil. Trans. 1803.

1803. Thomas Young.

"Experiments and Calculations relative to Physical Optics." Phil. Trans. 1804.

1804. Samuel Vince.

"Observations on the Hypotheses which have been assumed to account for the cause of Gravitation from Mechanical Principles." Journal Book, Vol. XXXVIII, p. 334.

1805. William Hyde Wollaston.

"On the Force of Percussion." Phil. Trans. 1806.

1806. Sir Humphry Davy, Bart.

"On some Chemical Agencies of Electricity." Phil. Trans. 1807.

1807. Sir Humphry Davy, Bart.

"On some new Phenomena of Chemical Changes produced by Electricity, particularly the Decomposition of the fixed Alkalies, and the Exhibition of the new Substances which constitute their Bases." Phil. Trans. 1808.

1808. Sir Humphry Davy, Bart.

"An Account of some new Analytical Researches on the Nature of certain Bodies, particularly the Alkalies, Phosphorus, Sulphur, Carbonaceous Matter, and the Acids hitherto undecompounded; with some general Observations on Chemical Theory." Phil. Trans. 1809, pp. 39, 450.

1809. Sir Humphry Davy, Bart.

"On some new Electro-Chemical Researches on various objects, particularly the Metallic Bodies from the Alkalies and Earths; and on some Combinations of Hydrogen." Phil. Trans. 1810.

1810. Sir Humphry Davy, Bart.

"On some of the Combinations of Oxymuriatic Gas and Oxygen, and on the Chemical Relations of these Principles to Inflammable Bodies." Phil. Trans. 1811.

1811. Sir Humphry Davy, Bart. (?)

1812. William Hyde Wollaston.

"On the Elementary Particles of certain Crystals." Phil. Trans. 1813.

1813. William Thomas Brande.

"On some new Electro-Chemical Phenomena." Phil. Trans. 1814.

1814 to 1818. No record.

1819. William Thomas Brande.

"On the Composition and Analysis of the inflammable Gaseous Compounds resulting from the destructive Distillation of Coal and Oil; with some Remarks on their relative heating and illuminating power." Phil. Trans. 1820.

1820. Captain Henry Kater.

"On the best kind of Steel, and form, for a Compass Needle." Phil. Trans. 1821.

1821. Captain Edward Sabine.

"An Account of Experiments to determine the Amount of the Dip of the Magnetic Needle in London, in August 1821; with Remarks on the Instruments which are usually employed in such determinations." Phil. Trans. 1822.

1822. No record.

1823. John F. W. Herschel.

"On certain Motions produced in Fluid Conductors when transmitting the Electric Current." Phil. Trans. 1824.

1824 to 1825. No record.

1826. Sir Humphry Davy, Bart.

"On the Relations of Electrical and Chemical Changes." Phil. Trans. 1826.

1827. George Pearson.

"Researches to discover the Faculties of Pulmonary Absorption with respect to Charcoal." Journal Book, Vol. XLV. p. 201.

1828. William Hyde Wollaston.

"On a Method of rendering Platina malleable." Phil. Trans. 1829.

1829. Michael Faraday.

"On the Manufacture of Glass for Optical Purposes." Phil. Trans. 1830.

1830. No record.

1831. No record.

1832. Michael Faraday.

"Experimental Researches in Electricity; Second Series."
Abstracts of Papers, Vol. III. p. 95.

1833. Samuel Hunter Christie.

"Experimental Determination of the Laws of Magneto-Electric Induction in different masses of the same metal, and of its intensity in different metals."
Abstracts of Papers, Vol. III. p. 177.

1834. Not appointed.

1835. Charles Lyell.

"On the Proofs of a gradual Rising of the Land in certain parts of Sweden." Phil. Trans. 1835.

1836. John William Lubbock.

"On the Tides of the Port of London." Phil. Trans. 1836.

1837. William Henry Fox Talbot.

"Further Observations on the Optical Phenomena of Crystals."
Phil. Trans. 1837.

1838. James Ivory.

"On the Theory of the Astronomical Refractions."
Phil. Trans. 1838.

1839. William Snow Harris.

"Inquiries concerning the Elementary Laws of Electricity."
Phil. Trans. 1839.

1840. George Biddell Airy.

"On the Theoretical Explanation of an apparent new Polarity of Light." Phil. Trans. 1840.

1841. George Newport.

"On the Organs of Reproduction and the Development of the Myriapoda." Phil. Trans. 1841.

1842. James David Forbes.

"On the Transparency of the Atmosphere and the Law of Extinction of the Solar Rays in passing through it."
Phil. Trans. 1842.

1843. Charles Wheatstone.

"An Account of several new Instruments and Processes for determining the Constants of a Voltaic Circuit."
Phil. Trans. 1843.

1844. Richard Owen.

"A Description of certain Belemnites, preserved, with a great proportion of their soft parts, in the Oxford Clay, at Christian-Malford, Wilts." Phil. Trans. 1844.

1845. Charles Giles Bridle Daubeny.

"Memoir on the Rotation of Crops, and on the Quantity of Inorganic Matters abstracted from the Soil by various Plants under different circumstances." Phil. Trans. 1845.

1846. James David Forbes.

"Illustrations of the Viscous Theory of Glacier Motion." Phil. Trans. 1846.

1847. William Robert Grove.

"On certain Phenomena of Voltaic Ignition and the Decomposition of Water into its constituent Gases by Heat." Phil. Trans. 1847.

1848. Rev. William Whewell.

"Researches on the Tides." Thirteenth Series. On the Tides of the Pacific, and on the Diurnal Inequality. Phil. Trans. 1848.

1849. Michael Faraday.

"Experimental Researches in Electricity." Twenty-Second Series. Phil. Trans. 1849.

1850. Thomas Graham.

"On the Diffusion of Liquids." Phil. Trans. 1850.

1851. Michael Faraday.

"Experimental Researches in Electricity." Twenty-Fourth Series. Phil. Trans. 1851.

1852. Charles Wheatstone.

"Contributions to the Physiology of Vision. Part II. On some remarkable and hitherto unobserved Phenomena of Binocular Vision (continued)." Phil. Trans. 1852.

1853. Col. Edward Sabine.

"On the Influence of the Moon on the Magnetic Declination at Toronto, St. Helena, and Hobarton." Phil. Trans. 1853.

1854. Thomas Graham.

"On Osmotic Force." Phil. Trans. 1854.

1855. John Tyndall.

"On the Nature of the Force by which Bodies are repelled from the Poles of a Magnet; to which is prefixed an account of some experiments on Molecular Influences." Phil. Trans. 1855.

1856. William Thomson.

"On the Electrodynamic Qualities of Metals." Phil. Trans. 1856.

1857. Michael Faraday.

"Experimental Relations of Gold (and other metals) to Light." Phil. Trans. 1857.

1858. John Peter Gassiot.

"On the Stratifications and Dark Band in Electrical Discharges as observed in Torricellian Vacua." Phil. Trans. 1858.

1859. Edward Frankland.

"Researches on Organo-metallic Bodies." Fourth Memoir. Phil. Trans. 1859.

1860. William Fairbairn.

"Experimental Researches to determine the Density of Steam at different Temperatures, and to determine the Law of Superheated Steam." Phil. Trans. 1860.

1861. John Tyndall.

"On the Absorption and Radiation of Heat by Gases and Vapours, and on the Physical Connection of Radiation, Absorption and Conduction." Phil. Trans. 1861.

1862. Warren De La Rue.

"On the Total Solar Eclipse of July 18, 1860, observed at Rivabellosa, near Miranda de Ebro, in Spain." Phil. Trans. 1862.

1863. Henry Clifton Sorby.

"On the Direct Correlation of Mechanical and Chemical Forces." Proceedings, Vol. 12, 1863.

1864. John Tyndall.

"Contributions to Molecular Physics: being the Fifth Memoir of Researches on Radiant Heat." Phil. Trans. 1864.

1865. Henry Enfield Roscoe.

"On a Method of Meteorological Registration of the Chemical Action of Total Daylight." Phil. Trans. 1865.

1866. James Clerk Maxwell.

"On the Viscosity or Internal Friction of Air and other Gases." Phil. Trans. 1866.

1867. Frederick Augustus Abel.

"Researches on Gun-Cotton. (Second Memoir.) On the Stability of Gun-Cotton." Phil. Trans. 1867.

1868. Henry Enfield Roscoe.

"Researches on Vanadium." Phil. Trans. 1868.

1869. Thomas Andrews.

"The Continuity of the Gaseous and Liquid States of Matter." Phil. Trans. 1869.

1870. John William Dawson.

"On the Pre-Carboniferous Flora of North Eastern America, and more especially on that of the Erian (Devonian) Period." Proceedings, Vol. 18.

1871. Charles William Siemens.

"On the Increase of Electrical Resistance in Conductors with Rise of Temperature, and its Application to the Measure of Ordinary and Furnace Temperatures: also on a simple Method of measuring Electrical Resistances." Proceedings, Vol. 19.

1872. William Kitchen Parker.

"On the Structure and Development of the Skull of the Salmon (*Salmo salar*, L.)." Proceedings, Vol. 20.

1873. Earl of Rosse.

"On the Radiation of Heat from the Moon, the Law of its Absorption by our Atmosphere, and its variation in Amount with her Phases." Proceedings, Vol. 21.

1874. J. Norman Lockyer.

"Researches in Spectrum Analysis in connection with the Spectrum of the Sun." Part III. Phil. Trans. 1874.

1875. William Grylls Adams.

"On the Forms of Equipotential Curves and Surfaces and on Lines of Flow." Proceedings, Vol. 24.

1876. Thomas Andrews.

"On the Gaseous State of Matter." Proceedings, Vol. 24.

1877. William Crawford Williamson.

"On the Organization of the Fossil Plants of the Coal Measures." Part IX. Phil. Trans. 1878.

1878. William Crookes.

"On Repulsion resulting from Radiation. Part V." Phil. Trans. 1878.

1879. William Crookes.

"On the Illumination of Lines of Molecular Pressure and the Trajectory of Molecules." Phil. Trans. 1879.

1880. Captain William de W. Abney.

"On the Photographic Method of Mapping the least refrangible end of the Solar Spectrum." Phil. Trans. 1880.

1881. John Tyndall.

"Action of free Molecules on Radiant Heat, and its conversion thereby into sound." Phil. Trans. 1882.

1882. Heinrich Debus.

"On the Chemical Theory of Gunpowder." Phil. Trans. 1882.

1883. William Crookes.

"On Radiant Matter Spectroscopy: the Detection and wide Distribution of Yttrium." Phil. Trans. 1883.

1884. Arthur Schuster.

"Experiments on the Discharge of Electricity through Gases. Sketch of a Theory." Proceedings, Vol. 37.

1885. William Huggins.

"On the Corona of the Sun." Proceedings, Vol. 39.

1886. Captain William de W. Abney and Major-General Edward Robert Festing.
"Colour Photometry." Phil. Trans. 1886.

1887. Joseph John Thomson.
"On the Dissociation of some Gases by the Electric Discharge." Proceedings, Vol. 42 (Abstract).

1888. J. Norman Lockyer.
"Suggestions on the Classification of the various Species of Heavenly Bodies. A Report to the Solar Physics Committee." Proceedings, Vol. 44.

1889. Arthur William Rücker and Thomas Edward Thorpe.
"A Magnetic Survey of the British Isles for the Epoch January 1, 1886." Phil. Trans. A, Vol. 181.

1890. Arthur Schuster.
"The Discharge of Electricity through Gases. Preliminary Communication." Proceedings, Vol. 47.

1891. George Howard Darwin.
"On Tidal Prediction." Phil. Trans. A, Vol. 182.

1892. James Thomson.
"On the Grand Currents of Atmospheric Circulation." Phil. Trans. A, Vol. 183.

1893. Harold B. Dixon.
"The Rate of Explosion in Gases." Phil. Trans. A, Vol. 184.

1894. Thomas Edward Thorpe and J. W. Rodger.
"On the Relations between the Viscosity (internal friction) of Liquids and their Chemical Nature." Phil. Trans. A, Vol. 185.

1895. A. G. Vernon Harcourt and William Esson.
"On the Laws of Connexion between the Conditions of a Chemical Change and its Amount. III. Further Researches on the Reaction of Hydrogen Dioxide and Hydrogen Iodide." Phil. Trans. A, 1895.

1896. William Chandler Roberts-Austen.
"On the Diffusion of Metals." Phil. Trans. A, 1896.

1897. Osborne Reynolds and W. H. Moorby.
"On the Mechanical Equivalent of Heat." Phil. Trans. A, Vol. 190

1898. William James Russell.
"Further Experiments on the Action exerted by certain Metals and other Bodies on a Photographic Plate." Proceedings, Vol. 63.

1899. James Alfred Ewing and W. Rosenhain.
"The Crystalline Structure of Metals." Phil. Trans. A, Vol. 193.

1900. William Augustus Tilden.
"On the Specific Heat of Metals and the Relation of Specific Heat to Atomic Weight." Phil. Trans. A, Vol. 194.

1901. James Dewar.
"The Nadir of Temperature and Allied Problems." Proceedings, Vol. 68 (Abstract).

1902. Lord Rayleigh.
"On the Law of the Pressure of Gases between 75 and 150 Millimetres of Mercury." Phil. Trans. A, Vol. 198.

1903. C. T. Heycock and F. H. Neville.
"On the Constitution of the Copper-tin Series of Alloys." Phil. Trans. A, Vol. 202.

1904. Ernest Rutherford.
"The Succession of Changes in Radio-active Bodies." Phil. Trans. A, Vol. 204.

1905. Horace T. Brown.
"The Reception and Utilisation of Energy by the Green Leaf." Proceedings, B, Vol. 76.

1906. John Milne.
"Recent Advances in Seismology." Proceedings, A, Vol. 77.

1907. Thomas Edward Thorpe.
"The Atomic Weight of Radium." Proceedings, A, Vol. 80.

1908. Charles H. Lees.
"The Effects of Temperature and Pressure on the Thermal Conductivities of Solids." Phil. Trans. A, Vol. 208.

1909. Sir Joseph Larmor.
"On the Statistical and Thermodynamical Relations of Radiant Energy." Proceedings, A, Vol. 83.

1910. John Henry Poynting and Guy Barlow.
"The Pressure of Light against the Source: the Recoil from Light." Proceedings, A, Vol. 83.

1911. Robert John Strutt.
"A Chemically-Active Modification of Nitrogen Produced by the Electric Discharge." Proceedings, A, Vol. 85.

1912. Hugh Longbourne Callendar.
"On the Variation of the Specific Heat of Water, with Experiments by a New Method." Phil. Trans. A, Vol. 212.

1913. Sir Joseph John Thomson.
"Rays of Positive Electricity." Proceedings, A, Vol. 89.

1914. Alfred Fowler.
"Series Lines in Spark Spectra." Phil. Trans. A, Vol. 214.

1915. William Henry Bragg.
"X-rays and Crystals." Phil. Trans. A, Vol. 215.

MEDAL AWARDS.

Copley Medal.

1731. Stephen Gray.
1732. Stephen Gray.
1734. John Theophilus Desaguliers.
1736. John Theophilus Desaguliers.
1737. John Belchier.
1738. James Valoue.
1739. Stephen Hales.
1740. Alexander Stuart.
1741. John Theophilus Desaguliers.
1742. Captain Christopher Middleton.
1743. Abraham Trembley.
1744. Henry Baker.
1745. Sir William Watson.
1746. Benjamin Robins.
1747. Gowin Knight.
1748. Rev. James Bradley.
1749. John Harrison.
1750. George Edwards.
1751. John Canton.
1752. Sir John Pringle.
1753. Benjamin Franklin.
1754. William Lewis.
1755. John Huxham.
1757. Lord Charles Cavendish.
1758. John Dollond.
1759. John Smeaton.
1760. Benjamin Wilson.
1764. John Canton.
1766. William Brownrigg.
Edward Delaval.
Hon. Henry Cavendish.
1767. John Ellis.
1768. Peter Woulfe.
1769. William Hewson.
1770. Sir William Hamilton.
1771. Matthew Raper.
1772. Joseph Priestley.
1773. John Walsh.
1775. Rev. Nevil Maskelyne.
1776. Captain James Cook.
1777. John Mudge.
1778. Charles Hutton.
1780. Rev. Samuel Vince.
1781. Sir William Herschel.
1782. Richard Kirwan.
1783. John Goodricke.
Thomas Hutchins.
1784. Edward Waring.
1785. Major-General William Roy.
1787. John Hunter.
1788. Sir Charles Blagden.
1789. William Morgan.
1791. James Rennell.
John Andrew De Luc.
1792. Benjamin Count Rumford.
1794. Professor Volta.
1795. Jesse Ramsden.
1796. George Attwood.
1798. Sir George Shuckburgh Evelyn.
Charles Hatchett.
1799. Rev. John Hellins.
1800. Edward Howard.
1801. Sir Astley Paston Cooper.
1802. William Hyde Wollaston.
1803. Richard Chenevix.
1804. Smithson Tennant.
1805. Sir Humphry Davy.
1806. Thomas Andrew Knight.
1807. Sir Everard Home.
1808. William Henry.
1809. Edward Troughton.
1811. Benjamin Collins Brodie.
1813. William Thomas Brande.
1814. James Ivory.
1815. David Brewster.
1817. Captain Henry Kater.
1818. Sir Robert Seppings.
1820. John Christian Oersted.
1821. Captain Edward Sabine.
John Frederick William Herschel.
1822. Rev. William Buckland.
1823. John Pond.
1824. John Brinkley, Bishop of Cloyne.
1825. François Arago.
Peter Barlow.
1826. Sir William South.
1827. William Prout.
Captain Henry Foster.

1831. George Biddell Airy.
1832. Michael Faraday.
Baron Simeon Denis Poisson.
1834. Giovanni Plana.
1835. William Snow Harris.
1836. Jöns Jacob Berzelius.
Francis Kiernan.
1837. Antoine C. Becquerel.
John Frederic Daniell.
1838. Karl Friedrich Gauss.
Michael Faraday.
1839. Robert Brown.
1840. Justus Liebig.
Jacques Charles François Sturm.
1841. George Simon Ohm.
1842. James MacCullagh.
1843. Jean Baptiste Dumas.
1844. Carlo Matteucci.
1845. Theodor Schwann.
1846. Urbain Jean Joseph Le Verrier.
1847. Sir John Frederick William Herschel.
1848. John Couch Adams.
1849. Sir Roderick Impey Murchison.
1850. Peter Andreas Hansen.
1851. Richard Owen.
1852. Baron Alexander von Humboldt.
1853. Heinrich Wilhelm Dove.
1854. Johannes Müller.
1855. Jean Bernard Lèon Foucault.
1856. Henry Milne-Edwards.
1857. Michel Eugène Chevreul.
1858. Sir Charles Lyell.
1859. Wilhelm Eduard Weber.
1860. Robert Wilhelm Bunsen.
1861. Louis Agassiz.
1862. Thomas Graham.
1863. Rev. Adam Sedgwick.
1864. Charles Darwin.
1865. Michel Chasles.
1866. Julius Plücker.
1867. Karl Ernst von Baer.
1868. Sir Charles Wheatstone.
1869. Henri Victor Regnault.
1870. James Prescott Joule.
1871. Julius Robert Mayer.
1872. Friedrich Wöhler.
1873. Hermann Ludwig Ferdinand Helmholtz.
1874. Louis Pasteur.
1875. August Wilhelm Hofmann.
1876. Claude Bernard.
1877. James Dwight Dana.
1878. Jean Baptiste Boussingault.
1879. Rudolph J. E. Clausius.
1880. James Joseph Sylvester.
1881. Karl Adolph Würtz.
1882. Arthur Cayley.
1883. Sir William Thomson.
1884. Carl Ludwig.
1885. August Kekulé.
1886. Franz Ernst Neumann.
1887. Sir Joseph Dalton Hooker.
1888. Thomas Henry Huxley.
1889. Rev. George Salmon.
1890. Simon Newcomb.
1891. Stanislao Cannizzaro.
1892. Rudolf Virchow.
1893. Sir George Gabriel Stokes.
1894. Edward Frankland.
1895. Carl Weierstrass.
1896. Carl Gegenbaur.
1897. Albert von Kölliker.
1898. Sir William Huggins.
1899. Lord Rayleigh.
1900. Marcellin Berthelot.
1901. Joseph Willard Gibbs.
1902. Lord Lister.
1903. Edward Suess.
1904. Sir William Crookes.
1905. Dmitri Ivanovitch Mendeleeff.
1906. Elias Metchnikoff.
1907. Albert A. Michelson.
1908. Alfred Russel Wallace.
1909. George William Hill.
1910. Sir Francis Galton.
1911. Sir George Darwin.
1912. Felix Klein.
1913. Sir Ray Lankester.
1914. Sir Joseph John Thomson.
1915. Ivan Petrovich Pavlov.

Rumford Medal.

1800. Benjamin Count Rumford.
1804. John Leslie.
1806. William Murdock.
1810. Etienne Louis Malus.
1814. William Charles Wells.
1816. Sir Humphry Davy.

1818. David Brewster.
1824. Augustin Jean Fresnel.
1832. John Frederic Daniell.
1834. Macedonio Melloni.
1838. James David Forbes.
1840. Jean Baptiste Biot.
1842. Henry Fox Talbot.
1846. Michael Faraday.
1848. Henri Victor Regnault.
1850. François Jean Dominique Arago.
1852. George Gabriel Stokes.
1854. Neil Arnott.
1856. Louis Pasteur.
1858. Jules Jamin.
1860. James Clerk Maxwell.
1862. Gustav Robert Kirchhoff.
1864. John Tyndall.
1866. Armand Hippolyte Louis Fizeau.
1868. Balfour Stewart.
1870. Alfred Olivier Des Cloizeaux.
1872. Anders Jonas Angström.
1874. Joseph Norman Lockyer.
1876. Pierre Jules César Janssen.
1878. Alfred Cornu.
1880. William Huggins.
1882. William de W. Abney.
1884. Tobias Robertus Thalén.
1886. Samuel Pierpont Langley.
1888. Pietro Tacchini.
1890. Heinrich Hertz.
1892. Nils C. Dunér.
1894. James Dewar.
1896. Philipp Lenard.
Wilhelm Conrad Röntgen.
1898. Oliver Joseph Lodge.
1900. Antoine Henri Becquerel.
1902. Charles Algernon Parsons.
1904. Ernest Rutherford.
1906. Hugh Longbourne Callendar.
1908. Hendrik Antoon Lorentz.
1910. Heinrich Rubens.
1912. Kamerlingh Onnes.
1914. Lord Rayleigh.

Royal Medals.

1826. John Dalton.
James Ivory.
1827. Sir Humphry Davy.
Friedrich Georg Wilhelm Struve.
1828. Johann Friedrich Encke.
William Hyde Wollaston.
1829. Charles Bell.
Eilhard Mitscherlich.
1830. David Brewster.
Antoine Jerome Balard.
1833. Auguste Pyrame De Candolle.
Sir John Frederick William Herschel.
1834. John William Lubbock.
Charles Lyell.
1835. Michael Faraday.
Sir William Rowan Hamilton.
1836. George Newport.
Sir John F. W. Herschel.
1837. Rev. William Whewell.
1838. Thomas Graham.
Henry Fox Talbot.
1839. James Ivory.
Dr. Martin Barry.
1840. Sir John F. W. Herschel.
Charles Wheatstone.
1841. Robert Kane.
Eaton Hodgkinson.
1842. William Bowman.
John Frederic Daniell.
1843. James David Forbes.
Charles Wheatstone.
1844. Thomas Andrews.
George Boole.
1845. George Biddell Airy.
Thomas Snow Beck.
1846. Michael Faraday.
Richard Owen.
1847. George Fownes.
William Robert Grove.
1848. Thomas Galloway.
Charles James Hargreave.
1849. Colonel Edward Sabine.
Gideon A. Mantell.
1850. Benjamin Collins Brodie.
Thomas Graham.
1851. Earl of Rosse.
George Newport.
1852. James Prescott Joule.

1852. Thomas Henry Huxley.
1853. Charles Darwin.
1854. August Wilhelm Hofmann.
Joseph Dalton Hooker.
1855. John Russel Hind.
John Obadiah Westwood.
1856. Sir John Richardson.
William Thomson.
1857. Edward Frankland.
John Lindley.
1858. Albany Hancock.
William Lassell.
1859. George Bentham.
Arthur Cayley.
1860. William Fairbairn.
Augustus Waller.
1861. William B. Carpenter.
James Joseph Sylvester.
1862. Rev. Thomas Romney Robinson.
Alexander William Williamson.
1863. Rev. Miles J. Berkeley.
John Peter Gassiot.
1864. Jacob Lockhart Clarke.
Warren De la Rue.
1865. Joseph Prestwich.
Archibald Smith.
1866. William Huggins.
William Kitchen Parker.
1867. John Bennet Lawes and Joseph Henry Gilbert.
Sir William Logan.
1868. Alfred Russel Wallace.
Rev. George Salmon.
1869. Sir Thomas Maclear.
Augustus Matthiessen.
1870. William Hallowes Miller.
Thomas Davidson.
1871. John Stenhouse.
George Busk.
1872. Thomas Anderson.
Henry John Carter.
1873. George James Allman.
Henry Enfield Roscoe.
1874. Henry Clifton Sorby.
William Crawford Williamson.
1875. William Crookes.
Thomas Oldham.
1876. William Froude.
Sir C. Wyville Thomson.
1877. Frederick Augustus Abel.
Oswald Heer.
1878. John Allan Broun.
Albert C. L. G. Günther.
1879. William Henry Perkin.
Andrew Crombie Ramsay.
1880. Joseph Lister.
Andrew Noble.
1881. Francis Maitland Balfour.
John Hewitt Jellett.
1882. William Henry Flower.
Lord Rayleigh.
1883. Thomas Archer Hirst.
J. S. Burdon Sanderson.
1884. George Howard Darwin.
Daniel Oliver.
1885. David Edward Hughes.
Edwin Ray Lankester.
1886. Francis Galton.
Peter Guthrie Tait.
1887. Colonel Alexander Ross Clarke.
Henry Nottidge Moseley.
1888. Baron Ferdinand von Mueller.
Osborne Reynolds.
1889. Walter Holbrook Gaskell.
Thomas Edward Thorpe.
1890. David Ferrier.
John Hopkinson.
1891. Charles Lapworth.
Arthur William Rücker.
1892. John Newport Langley.
Charles Pritchard.
1893. Arthur Schuster.
Harry Marshall Ward.
1894. Victor Alexander Haden Horsley.
Joseph John Thomson.
1895. James Alfred Ewing.
John Murray.
1896. Charles Vernon Boys.
Sir Archibald Geikie.
1897. Andrew Russell Forsyth.
Sir Richard Strachey.
1898. Walter Gardiner.
John Kerr.
1899. George Francis Fitzgerald.
William Carmichael McIntosh.
1900. Percy Alexander MacMahon.
Alfred Newton.
1901. William Edward Ayrton.
William Thomas Blanford.
1902. Horace Lamb.
Edward Albert Schäfer.
1903. Sir David Gill.

1903. Horace T. Brown.
1904. David Bruce.
William Burnside.
1905. John Henry Poynting.
Charles Scott Sherrington.
1906. Alfred George Greenhill.
Dukinfield Henry Scott.
1907. Ernest William Hobson.
Ramsay H. Traquair.
1908. John Milne.
Henry Head.
1909. Augustus E. H. Love.
Ronald Ross.
1910. John Joly.
Frederick Orpen Bower.
1911. William Maddock Bayliss.
George Chrystal.
1912. William Mitchinson Hicks.
Grafton Elliot Smith.
1913. Harold Baily Dixon.
Ernest Henry Starling.
1914. Ernest William Brown.
William Johnson Sollas.
1915. Joseph Larmor.
William Halse Rivers Rivers.

Davy Medal.

1877. Robert Wilhelm Bunsen.
Gustav Robert Kirchhoff.
1878. Louis Paul Cailletet.
Raoul Pictet.
1879. Paul Emile Lecoq de Boisbaudran.
1880. Charles Friedel.
1881. Adolf Baeyer.
1882. Dimitri Ivanovitch Mendeleeff.
Lothar Meyer.
1883. Marcellin Berthelot.
Julius Thomsen.
1884. Adolph Wilhelm Hermann Kolbe.
1885. Jean Servais Stas.
1886. Jean Charles Galissard de Marignac.
1887. John A. R. Newlands.
1888. William Crookes.
1889. William Henry Perkin.
1890. Emil Fischer.
1891. Victor Meyer.
1892. François Marie Raoult.
1893. J. H. van't Hoff.
J. A. Le Bel.
1894. Per Theodor Cleve.
1895. William Ramsay.
1896. Henri Moissan.
1897. John Henry Gladstone.
1898. Johannes Wislicenus.
1899. Edward Schunck.
1900. Guglielmo Koerner.
1901. George Downing Liveing.
1902. Svante August Arrhenius.
1903. Pierre Curie and Madame Curie.
1904. William Henry Perkin, Jun.
1905. Albert Ladenburg.
1906. Rudolf Fittig.
1907. Edward W. Morley.
1908. William Augustus Tilden.
1909. Sir James Dewar.
1910. Theodore W. Richards.
1911. Henry Edward Armstrong.
1912. Otto Wallach.
1913. Raphael Meldola.
1914. William Jackson Pope.
1915. Paul Sabatier.

Darwin Medal.

1890. Alfred Russel Wallace.
1892. Sir Joseph Dalton Hooker.
1894. Thomas Henry Huxley.
1896. Giovanni Battista Grassi.
1898. Karl Pearson.
1900. Ernst Haeckel.
1902. Francis Galton.
1904. William Bateson.
1906. Hugo de Vries.
1908. August Weismann.
1910. Roland Trimen.
1912. Francis Darwin.
1914. Edward Bagnall Poulton.

Buchanan Medal.

1897. Sir John Simon.
1902. Sydney A. Monckton Copeman.
1907. William Henry Power.
1912. William C. Gorgas.

Sylvester Medal.

1901. Henri Poincaré.
1904. Georg Cantor.
1907. Wilhelm Wirtinger.
1910. Henry Frederick Baker.
1913. James Whitbread Lee Glaisher.

Hughes Medal.

1902. Joseph John Thomson.
1903. Wilhelm Hittorf.
1904. Sir Joseph Wilson Swan.
1905. Augusto Righi.
1906. Mrs. Hertha Ayrton.
1907. Ernest Howard Griffiths.
1908. Eugen Goldstein.
1909. Richard Tetley Glazebrook.
1910. John Ambrose Fleming.
1911. Charles Thomson Rees Wilson.
1912. William Duddell.
1913. A. Graham Bell.
1914. John S. Townsend.
1915. Paul Langevin.

AWARD OF MEDALS, 1915.

The Copley medal to Prof. Ivan Petrovich Pavlov on the ground of his investigations in the Physiology of Digestion and of the higher centres of the Nervous System.

A Royal medal to Sir Joseph Larmor on the ground of his numerous and important contributions to Mathematical and Physical Science.

A Royal medal to Dr. W. H. Rivers on the ground of his important contributions to Ethnography and Ethnology.

The Davy medal to Prof. Paul Sabatier on the ground of his researches on Contact Action and the application of finely divided metals as Catalytic Agents.

The Hughes medal to Prof. Paul Langevin on the ground of his important contributions to, and pre-eminent position in, Electrical Science.

STUDENTSHIPS.

The Mackinnon Studentships are held by—

Mr. G. Matthai, of Emmanuel College, Cambridge, for a research on the Comparative Anatomy of the Madreporaria (second year).

Mr. A. E. Oxley, of Trinity College, Cambridge, for a continuation of his researches on the Influence of Molecular Constitution and Temperature on Magnetic Susceptibility (second year).

The Tyndall Studentship (second year) is held by Mr. J. I. Graham for an Investigation into the Spontaneous Combustion of Coal, with special reference to Gob-fires.

GUNNING FUND (1913).

The three years' accumulations of income accrued since 1910 have been allocated to Prof. J. P. Hill in aid of his Expedition to South America to collect material for the study of the Edentata and primitive Marsupials of America.

SORBY RESEARCH FELLOWSHIP (1914).

The Sorby Research Fellowship is held by Mr. B. Lightfoot, for a study of the Yorkshire Coalfield.

ANNIVERSARY MEETING.

1915.

On Tuesday, November 30, the Anniversary Meeting of the Society was held in the apartments in Burlington House.

SIR WILLIAM CROOKES, O.M., D.SC., President, in the Chair.

The Report of the Auditors was presented as follows:—

"During the year the total Ordinary Income on General Purposes Account, including the Treasury Grant of £1,000 for Publications, amounted to £7,536 9*s.* 6*d.*

"The total Ordinary Expenditure on General Purposes Account amounted to £6,570 1*s.*, and there was therefore an excess of Income over Expenditure of £966 8*s.* 6*d.*

"Extraordinary Income included a bequest by the late Mr. W. E. Darwin, amounting to £1,485, and a Special Grant from the Government Grant Committee towards Publications by other Societies of £500.

"After deducting the expenditure in connection with the Catalogue of Scientific Papers, there was an excess of Extraordinary Income over Expenditure of £1,785 16*s.* 1*d.*

"The Trust Funds show balances in the Bank on Current and Deposit Accounts amounting to £8,477 7*s.* 7*d.*

"The accounts of the National Physical Laboratory, the Government Grant Committee, and the International Catalogue are separately audited by Mr. W. B. Keen, Chartered Accountant."

The thanks of the Society were voted to the Treasurer and Auditors.

The Secretary read the lists of Fellows elected and deceased since the last Anniversary. (*See* p. 48.)

The following Report of the Council, which had been previously distributed to the Fellows, having been taken as read, was, on the motion of the President, received:—

REPORT OF COUNCIL, 1915.

When the Council re-assembled after the outbreak of war, the military operations had assumed a character presenting many scientific problems for solution.

The Council, accordingly, on November 5, 1914, passed the following resolution:—

> "That the following be appointed a Committee to organize assistance to the Government in conducting or suggesting scientific investigations in relation to the war, the Committee to have power to add to their number; and to appoint Sub-Committees not necessarily restricted to Fellows of the Society:—Sir William Crookes, Prof. Schuster, Mr. W. Duddell, Sir Alfred Ewing, Dr. R. T. Glazebrook, Admiral Sir Henry Jackson, Sir Oliver Lodge, Sir William Ramsay, Lord Rayleigh, Sir Edward Thorpe."

The Government departments chiefly concerned were made acquainted with the appointment of this Committee, and with its readiness to afford them scientific assistance.

The Admiralty, and somewhat later the War Office and the Board of Trade, responded to this offer, and a considerable variety of questions were submitted for solution. In order to deal with these problems effectively, the War Committee above referred to nominated Sub-Committees appropriate to the main branches of science concerned. On January 21, 1915, an additional Committee (the War Industries Committee) was appointed by the Council to take over questions arising out of the war and affecting the industries of the country referred to the Society by Government departments. This Committee undertook the duty of advising the Government on the question of the supply of drugs and other chemicals for which the country had hitherto been dependent on foreign supply. For this and other purposes the chemical laboratories of the principal scientific institutions of the country were asked to carry out the necessary work, and the Council have pleasure in acknowledging the help they have received from these institutions.

After a short experience the Council came to the conclusion that they themselves should be directly responsible for the co-ordination and control of the war work, and accordingly, in June last the original Committee was discharged, and Council was constituted a General War Committee, the original Sub-Committees being re-nominated as four sectional Committees. The functions of the War Industries Committee were transferred to the sectional Committee for Chemistry. These Sectional Committees have met regularly throughout the year, and have discussed and investigated many important questions, either submitted by Government departments or initiated by themselves. They have been placed in direct communication with the departments of Government concerned, and in each case representatives of the principal war departments have been appointed to serve on the Committee.

In the meantime several departments of the Government have themselves appointed Advisory Scientific Committees, upon two of which the Council were requested to nominate members, while Fellows of the Society individually have been invited to serve upon others. Thus the Admiralty, the Board of Trade, and the Ministry of Munitions have all appointed Committees or Boards to advise them upon general or special subjects connected with the prosecution of the war.

It has been the great desire of the Council to avoid the overlapping of scientific effort, and consequent waste of energy, in the conduct of the war; and in making the appointments to one of the Committees named above, the Council requested the Minister to receive a representation from the Society with regard to a better co-ordination of the various bodies rendering scientific advice to the Government. A memorandum on this subject was prepared for presentation to the Government. The Council approved the memorandum, but delayed its presentation until it had been submitted to two Fellows of the Society in close touch with the War Departments of Government. At an interview with them it was represented that the Admiralty was preparing a scheme which would substantially meet the views of the Society, and the presentation of the memorandum was accordingly withheld.

As a consequence of representations made to the President by a number of the Chemists, Fellows of the Society, the Council drew up a Memorial to the Prime Minister calling attention to the urgent necessity of establishing a closer relationship than has hitherto existed between scientific research and the industries of this country, and requested him to receive a deputation on the subject. The Prime Minister's other occupations prevented him from receiving the deputation himself; but at his request the President of the Board of Trade accepted the proposal, and a deputation composed of delegates of the Royal Society and the Chemical Society waited upon him accordingly on May 6, when the subject of the Memorial was fully discussed. The President of the Board of Education was present at the interview, and announced that he had a scheme in preparation for presentation to Parliament which he hoped would, to a large extent, meet the views of the deputation. This scheme in its final form is now issued as a Parliamentary Paper. A copy of the Council's Memorial is appended to this report.

War Register.

In June last the Council decided to form a Register of those who are willing and able to aid in meeting the demands of the Government for scientific help. A circular letter inviting particulars of service proffered was sent to all the Fellows and met with a large response from them and from others to whom the project was made known. The Register thus formed has been placed at the disposition of Government Departments for reference. Arising out of this Register, and based to a large extent upon it, a classified list has been prepared with the aid of other Scientific Societies, Universities and Colleges, of the men of science in the departments of Physics, Engineering and Chemistry who are serving in any capacity in the Forces of the Crown. Such a list is difficult to make complete, but a preliminary instalment has already been printed and placed in the hands of the Naval and Military Departments of Government, which have expressed appreciation of its usefulness; and a second continuation of the list is in preparation. The Council are greatly indebted to Dr. W. H. Eccles, one of the Secretaries of the Physical Society, for his help in compiling these lists, and owe their thanks also to the Societies and Universities which have assisted so materially with the information on which the lists are based.

Joint Action with other Societies.

A Committee has been appointed to consider the advisability of entering into communication with technical and other scientific societies, with a view to establishing a permanent Board for the discussion of questions in which joint action seems desirable. This Committee will submit the draft of a scheme to the Council, so as to put them in a better position to arrive at a definite conclusion on the subject.

Royal Society Catalogue of Scientific Papers.

The position of the Royal Society Catalogue of Scientific Papers has caused the Council much anxiety, having regard to the facts that the special fund provided for that work by outside contributions, amounting to £20,000, was exhausted at the end of 1913, and that the preparation of the Catalogue has since that date been carried on entirely at the Society's expense. Further delays have occurred in the printing; the Syndics of the Cambridge University Press, having been compelled to reduce the rate of printing, so that at the most only two sheets a week can be furnished instead of four, as estimated in the last Report of Council.

The total number of title slips comprised in the Authors' Catalogue amounts to about 443,175. Two volumes have now been published, as follows:—

Volume xiii (A and B), 1,049 pages, 51,770 entries.
Volume xiv (C to Fittig), 1,024 pages, 57,055 entries.

Volume xiii was issued on June 18, 1914; Volume xiv was issued on February 12, 1915.

The remaining entries are estimated to occupy five more volumes, namely, four volumes of about 1,000 pages or 125 sheets each, and 1 volume of about 765 pages.

Even if the Press continues to work at the expected rate of two sheets per week there is now no likelihood of the final volume being produced until the middle of 1921.

The Council have to report that Dr. McLeod has been obliged to relinquish his duties as Director of the Catalogue on the grounds of ill-health, and in recognition of their sense of the devotion and industry he brought to the laborious task entrusted to him, they have designated him Honorary Director of the Catalogue.

The remaining work of the Catalogue will be carried on under the supervision of the Chairman of the Committee without any new appointment to replace the Director. Owing to the delay in the number of sheets set up by the Press and the consequent reduction in the amount of proof reading, it has been found possible to make further reductions in the staff, which now consists of two seniors and two juniors.

The Council desire to acknowledge their indebtedness to Prof. Silvanus Thompson, Chairman of the Catalogue Committee, who has devoted much time and attention to stimulating and supervising the work of the preparation and publication of the Catalogue.

PUBLICATIONS.

During the last year 17 papers have been published in the Mathematical and Physical Section, and 9 in the Biological Section, of the "Philosophical Transactions." These two sections contain together 1,108 pages of letterpress and 33 plates.

Of the "Proceedings" 12 numbers have been issued in Series A and 8 in Series B. These contain in Series A (Mathematical and Physical) 1,150 pages and 8 plates; and in Series B (Biological) 456 pages and 7 plates.

STATUTES.

The Council have had under consideration some of the Statutes of the Society, and have made certain amendments embodying the Regulations adopted last year relating to Candidates for the Fellowship and in other minor particulars. The principal alterations to Chapter I. provide that (*a*) either of the two Secretaries shall have power to sign a certificate on behalf of a Fellow not resident in the United Kingdom on receiving a request from him; (*b*) of the Fellows who sign a certificate from personal knowledge of the Candidate, two shall undertake the responsibility of acting as proposer and seconder respectively, in so far that all communications on the subject of the candidature shall take place with one of these, and the proposer shall be responsible for informing the Candidate of the Obligation to be subscribed, and of the payments to be made to the Society before he can be admitted a Fellow; (*c*) any separate paper attached to the certificate, if intended to be printed for circulation to the Fellows, shall bear the signatures of six of the signatories to the certificate. Corrections of fact may be made upon the certificate after it has been sent in, if initialled by the proposer and seconder, and marked unmistakably as new matter.

A new Statute has been made which provides that a Special General Meeting of the Society shall be held before the Annual Meeting on Thursday preceding the Annual Meeting. At such Special Meeting the Report of the Council, having been previously circulated, shall be presented, and Fellows shall be invited to comment thereon. Matters not arising out of the Report shall not be discussed except by permission of the President. The Report shall be finally presented to the Society at the Anniversary Meeting, but no change shall be introduced after its first presentation other than such as the Council may think fit to adopt on consideration of the comments made at the Special General Meeting.

Such a Meeting was held on November 26, 1914, in pursuance of a resolution of Council, subsequently embodied in the Statutes.

In Chapter VI. relating to the procedure in the election of Council the Statute has been amended so as to give the Scrutators full responsibility for the examination of the balloting lists.

The revised Statutes will be printed in the next edition of the Year-book.

Government Grant for Scientific Investigations.

Pursuant to the recommendations of the Government Grant Committee, the President and Council this year have made grants amounting to £3,344. This amount includes a grant of £500 to the Royal Society for the purposes of the publications of the Society, the Treasury having agreed that during the continuance of the present crisis £500 per annum out of the grant may be devoted to the publication of the results of scientific research.

The sum of £403 9*s*., in addition to the unallotted balance of £96 11*s*. already in their hands from last year's grant, was reserved for administration by the Council to meet pressing demands upon the Fund which may arise before the next meeting of the Government Grant Committee.

The unexpended sums reclaimed during the year from holders of grants, under Regulation 27, and carried to the Reserve Fund Account, amount to £349 2*s*. 1*d*.

The unallotted balance of the grant, namely, £202 11*s*., has been carried to the Reserve Fund Account.

International Catalogue of Scientific Literature.

All the volumes of the first thirteen annual issues have been published with the exception of the following:—

Bacteriology of the 11th issue and Meteorology of the 12th issue are in the press; Mineralogy of the 12th issue and Astronomy, Geography and Palæontology of the 13th issue are in revise; Geology and Anatomy of the 12th issue, Meteorology, Mineralogy, Biology, Botany and Anthropology of the 13th issue are at the printers; Physiology and Bacteriology of the 12th and 13th issues, Chemistry, Geology and Anatomy of the 13th issues are in the experts' hands.

The war has retarded the publication of the Catalogue through shortage of labour at Messrs. Harrison & Sons, and through the large amount of Government printing which has had to be carried out by that firm, and on which the Government have demanded priority.

Since the outbreak of hostilities in August, 1914, no subscriptions have been received from Germany, Austria, Hungary, Poland, and

Belgium. Owing to the reduction of income thus occasioned, the Council have approved the grant of £1,100 from the Donation Fund to enable payments of an urgent character to be made. The other countries co-operating in the production of the Catalogue have been paying their subscriptions in the usual way. Germany and Austria continued the work of indexing their literature for the Catalogue till the end of March, 1915, at which date the German regional bureau was closed. The Austrian and Hungarian Bureaus are continuing their work as usual. The cards indexing German and Austrian literature for 1914 have been received at the Central Bureau.

The Council of the Royal Society referred to a Committee the question as to the attitude which should be taken with regard to the Catalogue. After receiving a report from the Committee the Council decided that in view of the impossibility, at the present time, of summoning the International Council to consult as to the future of the Catalogue, the Royal Society would guarantee the completion of the issues then in course of preparation, up to and including the thirteenth issue, but not further.

In consequence of this resolution the various Regional Bureaus have been notified that the collection of material for years subsequent to 1913 must be undertaken by them spontaneously and without prejudice, as it was impossible to guarantee the publication of issues later than the 13th.

Notwithstanding this notification, all the Bureaus have continued their work so that a large amount of material for the 14th annual issue is now at the Central Bureau.

At the present moment there are at the Central Bureau (omitting Zoology), 115,784 cards indexing the literature of 1914.

The volume of Zoology of the 14th issue was in a different position from the volumes relating to other sciences, because an agreement had been made with the Zoological Society for the amalgamations of Volume N of the International Catalogue with the Zoological Record.

The Zoological Society, however, accepted the financial responsibility of publishing this volume at the usual time, and without change of form.

The volume will be issued in the first instance only to those who subscribe for the Zoological Record in advance, but in provision for the event of a decision hereafter to print the whole of the 14th issue of the Catalogue, the Zoological Society will print a sufficient number of copies of the volume to enable it to be issued as part of the 14th annual issue of the Catalogue.

The material supplied by the Regional Bureaus has been used in preparing this volume.

Library.

During the year one new serial publication has been added to the Library, viz.:—

Journal of Agricultural Research, Washington.

Various works have been added to the Library by presentation and purchase. Among those presented are:—

Bayliss, W. M.—Principles of General Physiology.
Brazil, V.—La Défense contre l'Ophidisme.
Chappell, E.—Five-figure Mathematical Tables.
Darboux, G.—Leçons sur la Théme Générale des Surfaces. 2me Partie. 2me Édition.
Grassi, B.—Studi di uno Zoologo sulla Malaria. 2a Edizione.
Napier Tercentenary Celebration. Handbook of Exhibition of Relics, &c.
Scotia.—Report of Work carried out by the S.S. Scotia, 1913.

Magnetic Survey of the British Isles.

The reduction of the work of 1914 proceeded as far as possible during the winter of 1914–15.

The field work for 1915, started at the end of March. Owing to various circumstances, mainly connected with the War, the work has been subject to considerable interruption.

The survey of Ireland has been completed (44 stations), and 23 more points in England and Wales. Thus the mainland of Scotland, Ireland, and England has been completed (except for 6 more points in England and Wales, which it is hoped to complete this season). There remains to do the Hebrides (13 points), Isle of Man (2 points), and the Channel Isles (6 points). Throughout the work in Ireland occasion was taken to test the new portable variometer for measuring differences of horizontal force, designed by Mr. G. W. Walker, F.R.S., who is conducting the survey. The results obtained by this new method have been very gratifying and will be published soon.

The National Physical Laboratory.

The work of the Laboratory, like everything else in this country, has been greatly affected by the War. Some 25 per cent. of the Staff* have joined the colours, and a reorganisation of the work has been required in practically all divisions.

The financial results of the year ending March last were adversely affected by the war. Up to the end of June, 1914, the receipts for work done were considerably in excess of those for the corresponding three months of the preceding year; but during July, August and September there was a heavy fall amounting to some £2,100. After September, however, there was a considerable recovery and the result for the year is, as the accounts show, less unsatisfactory than was feared.

The Council has to report, with great regret, the death of Mr. E. B. Ellington, who was one of the representatives of the Institution of Mechanical Engineers on the General Board of the Laboratory, and gave assistance in many ways in connexion with the work of the Engineering Department. The recent death of Sir

* Of these Mr. D. Ewen was killed at Loos and Mr. Barwood has died of wounds received near Ypres.

Andrew Noble has deprived the Laboratory of the services of one of its earliest and most generous supporters. He was a member of Lord Rayleigh's Committee, which recommended the establishment of the Laboratory, and for many years took a close interest in the work and gave it his constant support. In recent years he rendered valuable assistance in the construction of the Lorenz apparatus. At the time of his death he was still a member of the Executive Committee.

Reference was made in last year's report to the services rendered to the Laboratory by Sir Arthur Rücker, who was also a member of Lord Rayleigh's Committee. His loss will be very keenly felt.

A large number of requests for advice and assistance have come from both the Admiralty and the War Office, and the Council have reason to believe that the help afforded has been of great value.

In addition to the aeronautics work, the time of a number of the senior members of the Staff has been given almost entirely to war research, and the subjects of enquiry have been very varied.

The Laboratory has been recognised by the Admiralty, and the Staff have been notified by Their Lordships that it is fully realized that in making provision for the requirements of the Royal Navy, they are doing their duty for their King and country, equally with those who have joined H.M. Forces afloat and ashore.

The Laboratory has, as usual, been consulted by various Government Departments, beside the Admiralty and War Office.

In some divisions, notably in the department of metallurgical chemistry, there has been a marked falling off in the test work; however, that in others it has been well maintained.

Dr. Harker has been appointed to represent the Laboratory on the Refractories Committee of the Institution of Gas Engineers. The need for research on refractories for furnace work of all kinds is well-known, and the Executive Committee hope the Laboratory may be able to assist the Institution in their enquiries.

International work has naturally not made much progress.

The work of Sir D. Gill's Committee, appointed in 1912 to consider the question of a Research into the Manufacture of Optical Glass, is now bearing fruit. The Treasury, on the motion of the Board of Trade, have promised grants of £1,500, £1,500 and £1,250 in this and the next two years; much of the necessary plant is at the Laboratory, and the experiments have commenced. The Laboratory has been in communication with the Institute of Chemistry with reference to this work.

The large impact testing machine erected by the help of the India Office and the Engineering Standards Committee, has, however, been completed, and the international research on methods of impact testing is in progress.

The Aeronautics Division and the Froude National Tank have both been fully occupied, though in consequence of the war the tests in the Tank on commercial forms of ships have been very few in number.

Details of the work of the several departments are given in the Director's Report.

The Report of the Metrology Division contains much interesting information, chiefly concerned with the maintenance of a high degree of accuracy in the standards of the Laboratory and in the method of testing instruments.

The Optical Division has been absorbed in test work called for by the requirements of the Admiralty and the War Office.

Dr. Stanton's report on the Engineering Department contains an interesting record of useful work and deals with a number of researches.

The researches in the Metallurgical Department have all made progress during the year, though the reduction of the Staff has greatly diminished the output. It is hoped that a further report on the Alloys Research work will be ready before the end of the year.

In the amount of test work for the India Office there has been, in consequence of the war, a marked falling off; the Staff, however, has been greatly reduced. Some of the investigatory test work in the Chemical Division, especially that on barium peroxide, for the Board of Trade, has been of great importance, while reference should also be made to a series of tests for Messrs. Chance Bros., on the weathering of optical glasses. The chemical work in connexion with the aeronautics researches has been of marked value, while much information has been obtained on the behaviour of bitumen and the binding materials used in road construction.

The Advisory Committee for the William Froude National Tank have again presented an interesting report, and the Council are indebted to them for their valuable help.

A list of publications of the Laboratory during the years 1913–14 and 1914–15 is given in the Report of the Executive Committee which contains full accounts of the researches carried out at the Laboratory during the year.

The accounts for the last financial year have been audited and presented to the Treasury in the usual form. They are appended to the Society's general accounts, and show a credit balance of £80 13*s.* 8*d.*

The result is much more satisfactory than the Executive Committee, in the earlier stages of the war, had dared to hope. It has only been reached by the exercise of the greatest care and economy in expenditure.

The Hon. Sir C. A. Parsons has recently placed a sum of £1,000 at the disposal of the Committee to be used at their discretion in additions to the equipment, and the Society will wish to thank him for his generosity, which will enable some marked deficiencies to be filled.

Last year steps were taken to bring the financial position of the Laboratory before the Government. A memorandum drawn up by the Executive Committee giving a statement of these reasons was presented to the Council in May, 1914, and a letter approved by the President and Council was sent, along with the memorandum, to Mr. Asquith in July, asking him to receive a deputation to lay before him reasons for relieving the Royal Society from the financial responsibility involved in their control of the Laboratory and for placing the Committee in a position to carry out efficiently work urgently required, by granting increased support from public funds. The receipt of the letter was acknowledged, but the outbreak of the war has prevented further action. The President and Council feel confident that, should the need arise, the Lords Commissioners of H.M. Treasury will take such measures as may be necessary to secure that the assistance of the Laboratory can be freely rendered in the development of the British industries after the war.

The late Sir Arthur Rücker.

The Council desire to place on record their sense of the eminent services rendered to Science and to the Society by the late Sir Arthur Rücker. Most of his important researches are published in the "Philosophical Transactions." During the five years he held a Secretaryship he conducted the affairs of the Society with exceptional skill, energy and tact; and he more especially took a prominent part in the institution of the National Physical Laboratory, and the organization of the International Catalogue of Scientific Literature.

New Council and Officers.

At the end of their last session the Council received with regret a request from Sir William Crookes that his name should not be put forward again for election to the office of President. Reluctant as the Society must be to lose the official services of so eminent a man of science, they cannot but respect his claims to be relieved from the burden of office; at the same time, they have the satisfaction of knowing that he is willing to remain an unofficial member of the Council for the ensuing year, and have accordingly included his name in the list put forward for ballot at the anniversary. As successor to Sir William Crookes in the office of President the Council have submitted the name of Sir Joseph J. Thomson for election by the Society.

The prolongation of the War has necessitated the continued absence of Sir John Bradford with the Expeditionary Force in France, and news has been received from him that there is no certain prospect of his return to civil life so long as the War lasts. In these circumstances, and particularly having regard to the important and urgent business which demands attention at the present time, the Council have thought it not to the interest of the Society that the Biological Secretaryship should remain in abeyance; and they have therefore nominated for election as Secretary for the ensuing year Mr. W. B. Hardy, V.P., who throughout the whole of the past year has discharged the duties of that office with great assiduity and success, for which the Council are very much indebted to him.

APPENDIX I.

Memorial to the Prime Minister on the Development of Chemical Industries.

The President and Council of the Royal Society have recently had under consideration the state of certain chemical industries in this country as revealed by the effects of the war. These industries are of the greatest national importance; they supply materials required in many different industries, including material required for war purposes, and on them depends the welfare of immense numbers of skilled workers. We are aware that the Government at an early stage foresaw the difficulties with which the country was likely to be confronted, and an Advisory Committee, formed under the auspices of the Board of Trade, has rendered and is rendering service in the desired direction.

As representatives of Science in this country we urge that the main causes of the comparatively backward state of certain industries are: (1) Failure to realize that modern industry, to be successful, must be based on scientific research; and (2) Want of more intimate association between the manufacturers and the workers in science. The Board of Trade has already given facilities for bringing together manufacturers, merchants and consumers, and the Treasury has recently formed a war trade department in such specialized branches of industry as we have had under consideration. It is essential for the future welfare of the country and for the permanent retention of such industries as may be established here as the result of the war, that some permanent central national organization should be created. We venture to hope that the Government will take the necessary steps, and we submit the case for your favourable consideration now because of its urgency, and because as a body of scientific workers we know that something more is required than temporary Committees. It is generally admitted that the foreign industries which have outstripped our own, and others which increasingly threaten us, have been built up as the result of about half-a-century's recognition by our competitors of the bearing of scientific research upon manufactures. If, as we believe, there has been neglect of this recognition here, the present crisis will no doubt lead to the development of more enlightened ideas, and we feel confident that a fresh departure in this country, under the lead of the Government would have a most important influence upon the future of the growing industries.

It is not for us to formulate any definite scheme for giving practical effect to the representations submitted in this Memorial, but there are certain special reasons for bringing forward the matter now. One of these reasons is that the Advisory Committee which has been already appointed by the Board of Trade contains the nucleus of such a permanent organization as in our opinion is essential for the future maintenance of our chemical industries in face of foreign competition. This Committee, which now comprises experts in several branches of chemistry bearing upon industry, might furnish material for the establishment of a larger standing Committee, of the nature of an Intelligence Department, serving the chemical industries in the same way that the Commercial Intelligence Department serves merchants and traders. A further reason for our urging the matter at the present time is that in view of the existing state of some of our industries, the chemists of this country generally consider that it is imperative on national grounds that all proposed new developments should be pushed forward with as little delay as possible. Should such an organization as we suggest be called into existence, we have reason to believe that the services and laboratories of the Chemical Fellows of the Society will be, so far as possible, at the disposal of the Government for the purpose of advising and, if necessary, of carrying out investigations which may be required by manufacturers for the solution of special problems.

We venture to hope that this Memorial may receive your favourable consideration for the reasons summarized below :—

(1) Many branches of industry dependent upon foreign chemical manufacture have been seriously affected by the war. The development of these industries in this country requires a very large amount of expert co-operation, and in view of the leeway which has to be made up a considerable increase in the number of research workers is necessary in order to hasten progress and to ensure the permanent retention of the new manufactures after the war.

(2) A Standing Advisory Committee comprising some of the foremost chemists in the country is already available as an organization for the service of the nation in the present emergency. We suggest that this Committee, if suitably strengthened, might form the nucleus of an organization such as we have in view.

(3) It is believed that if the Government would give official support to the formation of such a Chemical Committee under the auspices, say, of the Board of Trade, the confidence of manufacturers would be secured to an extent that has hitherto been unrealisable. With increased facilities of communication between manufacturers and expert advisers, it may confidently be anticipated that such advice would be more and more sought by those concerned, and our chemical industries benefited to a corresponding extent.

The President and Council of the Royal Society beg leave to assure you that the services of the Fellows are, as they always have been, at the disposal of the Government. In presenting this Memorial they are actuated solely by the desire to promote the interests of a group of large and important industries which have been particularly affected by the war. It is outside the province and beyond the resources of the Society as a purely scientific organization to undertake administrative control of any scheme which may be formulated with the immediate object of developing trade and manufacture, but realizing the vital importance of, and necessity for, a more intimate association between scientific experts and the leaders of industry, they urge most earnestly upon your Government to take the subject of this Memorial into serious consideration.

March 1, 1915.

BENEFACTIONS RECEIVED BY THE SOCIETY, 1915.

Under the will of the late Sir Arthur Church, F.R.S., a reversionary interest in forty-three £20 shares in the London and Westminster Bank, the income to be applied to the maintenance of the Society's Archives.

The late Lieut. H. G. J. Moseley, killed in action in the war, bequeathed all his estate to the Society, to be applied to the furtherance of experimental researches.

INCOME AND EXPENDITURE ACCOUNT.

GENERAL PURPOSES.

November 1st, 1914, to October 30th, 1915.

INCOME.

	£	s.	d.	£	s.	d.
To **Fees**—						
Annual Contributions, 355 at £3...	1,065	0	0			
" " 47 at £4...	188	0	0			
Composition Fees	120	0	0			
Fee Reduction Fund	505	0	0			
				1,878	0	0
" **Investments**—						
Rents	626	17	9			
Dividends and Interest	2,920	17	8			
Income Tax Recovered	212	7	1			
				3,760	2	6
" **Publications**—						
Sales:						
"Philosophical Transactions" ...	580	0	3			
"Proceedings"	386	3	11			
Herschel Reprint	12	10	0			
Miscellaneous	18	0	1			
	996	14	3			
Less Commission...	98	7	3			
	898	7	0			
Government Publication Grant ...	1,000	0	0			
				1,898	7	0

EXPENDITURE.

	£	s.	d.	£	s.	d.	£	s.	d.
By **Establishment Expenses**—									
Advertising				36	11	3			
Coal and Lighting				199	1	2			
Fire and other Insurance				93	2	7			
House Expenses				146	4	7			
Law Charges				6	10	2			
Miscellaneous Expenses				36	18	8			
Office Expenses				107	1	4			
Postages and Petty Charges ...				67	3	3			
Printing (Miscellaneous) and Stationery				230	9	5			
Salaries and Wages				2,782	6	6			
Receptions, and Meeting Expenses				66	12	6			
Taxes				44	14	0			
							3,816	15	5
" **Library**...							147	7	1
" **Publications**—									
"Philosophical Transactions"—									
Printing	650	10	6						
Paper	123	5	0						
Engraving and Lithography ...	260	9	9						
Distribution ...	128	14	4						
Binding	32	4	9						
				1,195	4	4			

	"Proceedings"—			
	Printing	676 5 1		
	Paper	119 10 9		
	Engraving and Lithography ...	149 11 3		
	Distribution ...	112 10 10		
	Binding	12 12 6		
			1,070 10 5	
	Year Book		72 3 9	
	Storage of Publications		35 0 0	
			2,372 18 6	
	Grants made during year for Extraneous Publications ...		233 0 0	
				2,605 18 6
				6,570 1 0
	,, **Balance, being excess of Ordinary Income over Expenditure**			966 8 6
£7,536 9 6				£7,536 9 6

EXTRAORDINARY INCOME AND EXPENDITURE ACCOUNT.

	£ s. d.		£ s. d.
To Bequest—Late Mr. W. E. Darwin	1,485 0 0	By Catalogue of Scientific Papers— Excess of Expenditure over Income	206 2 3
,, Special Grant by Government Grant Committee towards Publications by other Societies	500 0 0	,, Balance, being excess of Extraordinary Income over Expenditure	1,785 16 1
,, Sales of Charter Book, Record Book, Anniversary Publications, &c.	6 18 4		
	£1,991 18 4		£1,991 18 4

BALANCE SHEET.

GENERAL PURPOSES.

October 30th, 1915.

LIABILITIES.

	£ s. d.	£ s. d.
To Magnetic Survey Account—		
Balance from last Account	143 1 1	
Add Grant receivable from Caird Fund	200 0 0	
Grant receivable from Donation Fund	200 0 0	
Grant receivable from British Association ...	125 0 0	
	668 1 1	
Less Expenditure during Year	569 3 11	
		98 17 2
,, Publication Grant—		
Grants for Extraneous Publications not yet claimed, 1912	25 0 0	
,, ,, ,, 1914	90 0 0	
,, ,, ,, 1915	150 0 0	
		265 0 0
,, South African Government—		
Sale of Magnetic Reports		18 2 4
,, War Service Grant	250 0 0	
Less Expenditure during Year ...	87 4 8	
		162 15 4
,, Radium Research Fund—		
Balance from last Account ...	515 19 3	
Interest on Deposit...	11 7 9	
	527 7 0	
Less Expenditure during Year ...	100 0 0	
		427 7 0

ASSETS.

	£ s. d.	£ s. d.
By Cash at Bank, Current Account		98 17 2
,, Cash at Bank, Current Account		265 0 0
,, Cash at Bank, Current Account		18 2 4
,, Cash at Bank, Current Account		162 15 4
,, Cash at Bank, Deposit Account		427 7 0

,, Special Funds Account—			
Balance from last Account ...		677 8 1	
Interest on Deposit Account ...		9 5 2	
Sales of Antarctic Reports ...		0 16 1	
Grant from Home Office for Swine Fever Research		1,200 0 0	
		1,887 9 4	
	£ s. d.		
Less Expenditure—			
Sleeping Sickness	4 10 0		
Glassworkers' Cataract ...	100 0 0		
Swine Fever...	1,127 2 3		
		1,231 12 3	
			655 17 1
,, Capital Monies Temporarily Invested in Loan to International Catalogue Account, viz.:—			
Tyndall Fund		1,000 0 0	
McClean Legacy (part)		1,000 0 0	
Proceeds of Sale of India Stock		3,000 0 0	
			5,000 0 0
,, Income and Expenditure Account—			
Balance as per last Balance Sheet...		3,493 9 5	
Add Excess of Ordinary Income over Expenditure for the year		966 8 6	
Excess of Extraordinary Income over Expenditure for the year		1,785 16 1	
			6,245 14 0
			£12,873 12 11

,, Cash at Bank, Current Account ...		233 18 8	
,, do. Deposit ,, ...		421 18 5	
			655 17 1
Representing the following funds:—			
Antarctic		17 12 4	
Sleeping Sickness		389 2 11	
Glassworkers' Cataract		176 4 1	
Swine Fever		72 17 9	
		£655 17 1	
,, International Catalogue Account			7,500 0 0
Loan from the following sources:—			
Capital Monies as *per contra* ...		5,000 0 0	
Balance from General Purposes		2,500 0 0	
		£7,500 0 0	
,, Rents Receivable in arrear:—			
57, Basinghall Street		186 9 2	
Rayleigh Road, West Kensington		9 2 11	
,, International Catalogue Interest in arrear (£450 interest from May 5, 1914, not brought into account) ...		150 0 0	
,, Annual Contributions in arrear, ...		23 0 0	
,, Grants receivable (Magnetic Survey)		525 0 0	
			893 12 1
,, Cash at Bank, Current Account ...		2,836 10 3	
,, Cash in hand		15 11 8	
			2,852 1 11
			£12,873 12 11

CATALOGUE OF SCIENTIFIC PAPERS, 1884–1900, AND SUBJECT INDEX, 1801–1900.

Receipts.	£	s.	d.	Payments.	£	s.	d.
To Transfer from Handley Fund	183	11	2	By Salaries, Wages, and Office Expenses	889	13	5
,, Special Grant by Government Grant Committee ...	500	0	0				
,, Balance, amount overdrawn transferred to Extraordinary Income and Expenditure Account ...	206	2	3				
	£889	13	5		£889	13	5

ESTATES AND PROPERTY OF THE ROYAL SOCIETY.

GENERAL PURPOSES.

Estate at Mablethorpe, Lincolnshire, rent £75 per annum.

Ground Rent of House, No. 57, Basinghall Street, £380 per annum.

,, ,, of 23 houses in Rayleigh Road, West Kensington, £253 per annum.

Fee Farm Rent, near Lewes, Sussex, £19 4*s.* per annum.

Chancery Dividend. One-fourth annual interest on Balance of Bequest still in Court, amounting this year to £105 19*s.* 11*d.* (Stevenson Bequest).

£12,000 Mortgage Loan, 3½ per Cent., to the Duke of Norfolk.

£3,518 0*s.* 3*d.* 2½ per Cent. Consolidated Stock in Chancery, arising from sale of the Coleman Street Estate.

£220 7*s.* 6*d.* India 3½ per Cent. Stock.

£1,300 India 3 per Cent. Stock (Earl of Derby's Bequest).

£592 5*s.* 9*d.* Midland Railway Consolidated 2½ per Cent. Perpetual Guaranteed Preferential Stock (Stevenson Bequest).

£273 1*s.* 6*d.* Madras Railway "B" Annuity, repayable by Sinking Fund on April 1st, 1956.

£2,725 Great Northern Railway Consolidated 4 per Cent. Perpetual Preference Stock (Stevenson Bequest).

£14,908 London and North Western Railway Consolidated 4 per Cent. Guaranteed Stock.—{ £12,150 General Purposes. £2,758 Stevenson Bequest.

£5,000 " " " " Consolidated 4 per Cent. Preference Stock.

£5,818 North Eastern Railway 4 per Cent. Preference Stock.

£2,760 " " " Consolidated 4 per Cent. Guaranteed Stock (Stevenson Bequest).

£3,333 London and South Western Railway 4 per Cent. Consolidated Preferential Stock, 1881.

£4,000 Madras and Southern Mahratta Railway Company, Limited, 4 per Cent. Debenture Stock.

£4,182 5*s.* 2*d.* 2½ per Cent. Consolidated Stock.

£5,224 12*s.* 0*d.* New Zealand 3½ per Cent. Stock (Wolfe Bequest).

£1,650 New Zealand 4 per Cent. Stock.

£1,766 5*s.* 9*d.* Canada 4 per Cent. Stock.

£3,159 3*s.* 8*d.* Great Western Railway 4 per Cent. Debenture Stock (Wolfe Bequest).

£181 16*s.* 4*d.* " " " (General Purposes).

£1,500 Madras and Southern Mahratta Railway Company, Limited, Capital Stock.

£2,100 Caledonian Railway 4 per Cent. Consolidated Preference Stock.

£500 China Navigation Company, Limited, Ordinary Stock.

£700 Ontario and Quebec Railway 5 per Cent. Permanent Debenture Stock.

TRUST FUNDS. CASH ACCOUNTS, 1914–15.

BAKERIAN LECTURE AND COPLEY MEDAL FUNDS.

£1,666 13*s.* 4*d.* 2½ per Cent. Consolidated Stock. (Sir Joseph Copley's Gift.)
£403 9*s.* 8*d.* 2½ per Cent. Annuities.

	£	*s.*	*d.*		£	*s.*	*d.*
To Dividends, 2½ per Cent. Annuities	8	18	8	By Balance	0	3	3
,, Dividends—Sir J. Copley's Fund	36	18	0	,, Gold Medal—Sir J. J. Thomson	6	5	2
,, Income Tax recovered	2	17	0	,, Gift do.	41	0	4
,, Balance	2	15	1	,, Lecture Fee	4	0	0
	£51	8	9		£51	8	9

BRADY LIBRARY FUND.

£330 2½ per Cent. Consolidated Stock.

	£	*s.*	*d.*		£	*s.*	*d.*
To Balance	69	11	3	By Purchase of Books	0	10	10
,, Dividends...	7	6	4	,, Balance	76	15	11
,, Income Tax recovered	0	9	2				
	£77	6	9		£77	6	9

BUCHANAN MEDAL FUND.

£258 9*s.* 2*d.* Metropolitan 3 per Cent. Stock.

	£	*s.*	*d.*		£	*s.*	*d.*
To Balance	15	8	10	By Balance	22	14	9
,, Dividends	6	17	4				
,, Income Tax recovered...	0	8	7				
	£22	14	9		£22	14	9

CAIRD FUND.

£4,300 on Deposit at Bank.

	£ s. d.			£ s. d.
To Balance ...	4,545 17 8	By Grant for Seismological Work ...		200 0 0
,, Interest on Deposit Account ...	98 13 11	,, Balance ...	144 11 7	4,444 11 7
		,, ,, Deposit Account ...	4,300 0 0	
	£4,644 11 7			£4,644 11 7

CROONIAN LECTURE FUND.

£1,212 10*s*. 10*d*. War Loan 4½ per Cent. Stock (cost £1,206), and £120 Cash retained by Charity Commissioners out of proceeds of sale of Estate to meet increment value duty.

	£ s. d.		£ s. d.
To Balance ...	26 3 11	By Balance ...	61 2 2
,, Rent ...	23 6 9		
,, Interest ...	8 9 6		
,, Income Tax recovered ...	3 2 0		
	£61 2 2		£61 2 2

DARWIN MEDAL FUND.

£2,500 South Eastern Railway 4 per Cent. Perpetual Debenture Stock.
£560 South Eastern Railway 3 per Cent. Perpetual Debenture Stock.

	£ s. d.			£ s. d.
To Balance ...	637 11 4	By Gift—Prof. E. B. Poulton ...		100 0 0
,, Dividends ...	105 7 3	,, Medal ...		0 16 0
,, Interest on Deposit ...	5 9 5	,, Balance ...	404 1 6	654 1 6
,, Income Tax recovered ...	6 9 6	,, ,, Deposit Account ...	250 0 0	
	£754 17 6			£754 17 6

o

DAVY MEDAL FUND.

£36 0*s.* 11*d.* Madras Railway B Annuity, repayable by Sinking Fund on April 1st, 1956.

	£	*s.*	*d.*		£	*s.*	*d.*
To Balance ...	40	9	5	By Gold Medal—Prof. W. J. Pope ...	32	8	10
,, Dividends ...	24	18	1	,, Balance ...	34	7	0
,, Income Tax recovered...	1	8	4				
	£66	15	10		£66	15	10

DONATION FUND.

£5,030 Great Northern Railway Consolidated 4 per Cent. Perpetual Guaranteed Stock.
£1,861 6*s.* 8*d.* Great Northern Railway 3 per Cent. Debenture Stock. (Trevelyan Bequest.)
£2,000 on deposit at 5 per Cent. with Dewrance & Co. (Mr. John Dewrance's Gift).

	£	*s.*	*d.*		£	*s.*	*d.*
To Balance ...	1,054	1	7	By Grant, International Catalogue ...	1,100	0	0
,, Dividends ...	231	16	1	,, Balance ...	421	14	0
,, Interest on Deposit Accounts ...	99	9	8				
,, Income Tax recovered ...	14	4	9				
,, Transfer from Jodrell Fund ...	122	1	11				
	£1,521	14	0		£1,521	14	0

EMBOSSED SCIENTIFIC BOOKS FUND.

£1,550 New Zealand 3 per Cent. Stock.

	£	*s.*	*d.*		£	*s.*	*d.*
To Dividends ...	£41	3	6	By Payments to Committee ...	£41	3	6

FEE REDUCTION FUND.

£6,369 14*s.* 4*d.* Metropolitan 3½ per Cent. Stock.
£9,533 London and North Western Railway 3 per Cent. Perpetual Debenture Stock.

	£	s.	d.		£	s.	d.
To Balance ...	11	9	11	By Transfer to Royal Society Account ...	505	0	0
,, Dividends ...	455	1	5				
,, Income Tax recovered ...	28	3	9				
,, Balance ...	10	4	11				
	£505	0	0		£505	0	0

GASSIOT TRUST FUND.

£3,940 Italian Irrigation, Guaranteed 6 per Cent. (Cavour Canal) Bonds.
£32 5*s.* 1*d.* 2½ per Cent. Consolidated Stock.
£2,901 8*s.* 4*d.* Bank of England Stock.

	£	s.	d.		£	s.	d.
To Balance ...	2,500	1	3	By Purchase of £2,901 8*s.* 4*d.* Bank of England Stock	7,030	5	0
,, 3,980 Bonds drawn ...	4,616	16	0	,, Transfer to Gassiot Committee ...	547	17	9
,, Dividends ...	495	4	6	,, Balance ...	87	8	3
,, Interest on Deposit ...	28	8	6				
,, Income Tax recovered ...	25	0	9				
	£7,665	11	0		£7,665	11	0

GASSIOT COMMITTEE.

	£	s.	d.		£	s.	d.
To Balance ...	45	6	7	By Payments to Meteorological Committee ...	1,404	0	5
,, Gassiot Trust—Income for year ...	547	17	9	,, Balance ...	200	13	3
,, Paymaster - General — Grant for Eskdalemuir Observatory ...	1,000	0	0				
,, Rosse Fund—Income transferred ...	11	9	4				
	£1,604	13	8		£1,604	13	8

GORE FUND.

£2,478 14*s.* 4*d.* India $3\frac{1}{2}$ per Cent. Stock.

	£ s. d.			£ s. d.
To Balance ...	254 11 10	By Grant—Roscoff Marine Laboratory ...		60 0 0
„ Dividends	76 16 6	„ Balance ...	56 3 7	281 3 7
„ Interest on Deposit	4 19 0	„ „ Deposit Account ...	225 0 0	
„ Income Tax recovered ...	4 16 3			
	£341 3 7			£341 3 7

GUNNING FUND.

£1,081 5*s.* 6*d.* Nottingham Corporation 3 per Cent. Redeemable Stock.

	£ s. d.		£ s. d.
To Balance ...	16 3 4	By Balance ...	47 5 11
„ Dividends	29 6 7		
„ Income Tax recovered ..	1 16 0		
	£47 5 11		£47 5 11

HANDLEY FUND.

£4,798 Lancashire and Yorkshire Railway Consolidated 4 per Cent. Guaranteed Stock.

	£ s. d.		£ s. d.
To Dividends	172 18 6	By Transfer to Catalogue Account	183 11 2
„ Income Tax recovered ...	10 12 8		
	£183 11 2		£183 11 2

HUGHES MEDAL FUND.

£2,099 13*s*. 8*d*. Bath Corporation 3 per Cent. Redeemable Stock.
£2,094 6*s*. 6*d*. Local Loans 3 per Cent. Stock.

	£	s.	d.		£	s.	d.
To Balance ...	124	8	9	By Gold Medal—Prof. J. S. Townsend ...	20	14	8
„ Dividends	111	8	3	„ Gift— do.	104	8	1
„ Income Tax recovered	6	19	6	„ Balance ...	117	13	9
	£242	16	6		£242	16	6

JODRELL FUND.

£5,182 14*s*. 10*d*. 2½ per Cent. Consolidated Stock.

	£	s.	d.		£	s.	d.
To Dividends	114	18	6	By Transfer to Donation Fund ...	122	1	11
„ Income Tax recovered ...	7	3	5				
	£122	1	11		£122	1	11

JOULE MEMORIAL FUND.

£1,000 London, Brighton, and South Coast Railway Consolidated Guaranteed 5 per Cent. Stock.
£161 6*s*. 3*d*. 2½ per Cent. Annuities.

	£	s.	d.		£	s.	d.
To Balance ...	117	10	1	By Balance...	169	6	8
„ Dividends	49	1	5				
„ Income Tax recovered ...	2	15	2				
	£169	6	8		£169	6	8

KECK BEQUEST.

£960 Midland Railway 2½ per Cent. Debenture Stock.

	£	s.	d.		£	s.	d.
To Dividends	21	13	0	By Payment to Foreign Secretary	22	19	7
,, Income Tax recovered...	1	6	7				
	£22	19	7		£22	19	7

LAWRENCE FUND.

£4,055 9*s.* 3*d.* New South Wales 4 per Cent. Stock.

	£	s.	d.		£	s.	d.
To Dividends	143	12	8	By Balance	143	12	8
	£143	12	8		£143	12	8

MACKINNON STUDENTSHIP TRUST.

£4,482 6*s.* 0*d.* 2½ per Cent. Consolidated Stock.
£1,000 Metropolitan 2½ per Cent. Stock.
£5,991 London and North Western Railway 3 per Cent. Perpetual Debenture Stock.
The Trustees hold £3,334 London and North Western Railway 3 per Cent. Perpetual Debenture Stock, which will be transferred to the Society on the cessation of an annuity, and some small investments for sale.

	£	s.	d.		£	s.	d.
To Balance	413	10	8	By Grants	225	0	0
,, Dividends	283	6	0	,, Expenses	4	16	6
,, Income Tax recovered	17	10	11	,, Balance	484	11	1
	£714	7	7		£714	7	7

ROSSE FUND.

£119 18*s.* 6*d.* Bank of Ireland Stock.

	£	s.	d.		£	s.	d.
To Dividends	10	16	0	By Transfer to Gassiot Committee Account	11	9	4
,, Income Tax recovered...	0	13	4				
	£11	9	4		£11	9	4

RUMFORD FUND.

£2,367 2*s.* 6*d.* 2½ per Cent. Consolidated Stock.

	£	s.	d.		£	s.	d.
To Balance	117	16	6	By Gold Medal—Lord Rayleigh	58	16	2
,, Dividends	52	8	0	,, Gift do.	59	0	4
,, Income Tax recovered	3	5	6	,, Balance	55	13	6
	£173	10	0		£173	10	0

SCIENTIFIC RELIEF FUND.

£6,000 London and North Western Railway Consolidated 4 per Cent. Guaranteed Stock.
£4,340 South Eastern Railway 5 per Cent. Perpetual Debenture Stock.
£7,131 2*s*. 9*d*. Great Northern Railway 3 per Cent. Debenture Stock.

£126	Ditto	Ditto	Ditto (Sir J. D. Hooker's Gift, applicable as Income).
£68 17*s*. 3*d*.	Ditto	Ditto	Ditto (Investment of Income).

£737 10*s*. 0*d*. 2½ per Cent. Annuities (Investment of Income).

	£ s. d.	£ s. d.		£ s. d.
To Balance, Income	1,179 8 7		By Grants ...	1,015 0 0
Less Temporarily Invested ...	922 12 1		„ Expenses ...	0 15 0
		256 16 6		
„ Dividends		626 4 7		
„ Income Tax recovered ...		38 10 4		
„ Balance		94 3 7		
Viz., Income	828 8 6			
Less Temporarily Invested ...	922 12 1			
Overdraft	£94 3 7			
		£1,015 15 0		£1,015 15 0

SORBY RESEARCH FUND.

£15,503 16*s*. 6*d*. India 3½ per Cent. Stock.
£246 5*s*. 6*d*. 2½ per Cent. Consolidated Stock.

	£ s. d.		£ s. d.
To Balance ...	476 7 5	By Expenses ...	2 14 0
„ Dividends ...	485 18 4	„ Balance ...	989 19 9
„ Income Tax recovered ...	30 8 0		
	£992 13 9		£992 13 9

SYLVESTER MEDAL FUND.

£795 16*s.* 5*d.* Local Loans 3 per Cent. Stock.

	£	s.	d.		£	s.	d.
To Balance	23	14	3	By Balance	46	3	2
,, Dividends	21	2	10				
,, Income Tax recovered...	1	6	1				
	£46	3	2		£46	3	2

TRAVELLING EXPENSES FUND.

£1,504 19*s.* 6*d.* Bristol Corporation 3½ per Cent. Redeemable Stock.

	£	s.	d.		£	s.	d.
To Balance	135	11	2	By Balance	185	2	5
,, Dividends	46	12	10				
,, Income Tax recovered...	2	18	5				
	£185	2	5		£185	2	5

TYNDALL MINING RESEARCH FUND.

£1,010 1*s.* 0*d.* New Zealand 3½ per Cent. Stock.

	£	s.	d.		£	s.	d.
To Balance ...	39	17	6	By Studentship Grant ...	17	10	0
,, Dividends ...	31	6	0	,, Balance ...	55	12	7
,, Income Tax recovered ...	1	19	1				
	£73	2	7		£73	2	7

WINTRINGHAM FUND.

£1,200 2½ per Cent. Consolidated Stock.

	£	s.	d.		£	s.	d.
To Balance ...	29	16	5	By Payment to Foundling Hospital ...	29	16	5
,, Dividends ...	26	11	2	,, Balance ...	28	4	5
,, Income Tax recovered ...	1	13	3				
	£58	0	10		£58	0	10

BALANCE SHEET.

TRUST FUNDS.

October 30th, 1915.

Liabilities.	£	s.	d.
To Brady Library Fund	76	15	11
,, Buchanan Medal Fund	22	14	9
,, Caird Fund	4,444	11	7
,, Croonian Lecture Fund	61	2	2
,, Darwin Medal Fund	654	1	6
,, Davy Medal Fund	34	7	0
,, Donation Fund	421	14	0
,, Gassiot Trust Fund	87	8	3
,, Gassiot Committee	200	13	3
,, Gore Fund	281	3	7
,, Gunning Fund	47	5	11
,, Hughes Medal Fund	117	13	9
,, Joule Memorial Fund	169	6	8
,, Lawrence Fund	143	12	8
,, Mackinnon Studentship Fund	484	11	1
,, Rumford Fund	55	13	6
,, Sorby Research Fund	989	19	9
,, Sylvester Medal Fund	46	3	2
,, Travelling Expenses Fund	185	2	5
,, Tyndall Mining Research Fund	55	12	7
,, Wintringham Fund	28	4	5
	£8,607	17	11

We, the Auditors of the Treasurer's Accounts on the part of the Council, have examined the above Accounts and have found them correct. The Accounts of the National Physical Laboratory, the Government Grant Committee, and the International Catalogue are separately audited by Mr. W. B. Keen, Chartered Accountant.

D. H. SCOTT.
A. S. WOODWARD.
A. E. H. LOVE.

Assets.	£	s.	d.	£	s.	d.
By Cash at Bank on Current Accounts:—						
General Trusts Account...	3,796	11	2			
Less Scientific Relief overdrawn	94	3	7			
				3,702	7	7
,, Cash at Bank on Deposit:—						
General Trusts Account				4,775	0	0
,, Rent outstanding				23	6	9
,, Bakerian Lecture and Copley Medal Funds:— Amount Overdrawn				2	15	1
,, Fee Reduction Fund— Amount Overdrawn				10	4	11
,, Scientific Relief Fund— Amount Overdrawn				94	3	7
				£8,607	17	11

A. B. KEMPE, *Treasurer.*

We, the Auditors of the Treasurer's Accounts on the part of the Society, have examined the above Accounts and have found them correct. The Accounts of the National Physical Laboratory, the Government Grant Committee, and the International Catalogue are separately audited by Mr. W. B. Keen, Chartered Accountant.

A. STRAHAN.
JOHN M. THOMSON.
ALEX. PEDLER.

GOVERNMENT GRANT FOR SCIENTIFIC INVESTIGATIONS.

Income and Expenditure Account, April 1, 1914, to March 31, 1915.

This Fund is administered by the Royal Society, payments thereout being controlled by the specially constituted Government Grant Committee.

GENERAL FUND.

	£	s.	d.		£	s.	d.
To Parliamentary Grant ...	4,000	0	0	By Grants ...	4,384	0	0
,, Transfer from Reserve Fund ...	624	0	0	,, Salary, printing, postage, advertising, &c. ...	240	0	0
	£4,624	0	0		£4,624	0	0

RESERVE FUND.

	£	s.	d.		£	s.	d.
To Balance, April 1, 1914 ...	1,246	16	8	By Transfer to General Fund ...	624	0	0
,, Unexpended Balances called in ...	349	2	1	,, Assignment to Royal Society for expenses of Government Enquiries (Regulation 34) ...	70	0	0
,, Grants lapsed ...	20	0	0	,, Balance ...	1,021	6	11
,, Interest on Deposit ...	23	8	8				
,, Sale of apparatus ...	55	19	6				
,, Unexpended Balance of Administrative Expenses Assignment ...	20	0	0				
	£1,715	6	11		£1,715	6	11

NATIONAL PHYSICAL LABORATORY.

INCOME AND EXPENDITURE ACCOUNT FOR THE YEAR ENDED MARCH 31, 1915.

Income.

	£	s.	d.
To Treasury Grant for the year	7,000	0	0
,, Ditto Aeronautics, expended to March 31, 1915, (leaving £1,144 4*s*. 6*d*. carried forward) ...	7,474	10	8
,, William Froude National Tank— Subscriptions from Guarantors, &c.	1,552	0	0
,, Grants from Technical Societies	700	0	0
,, Charges for work done	22,777	9	0
,, Other Receipts	149	5	10
	£39,653	5	6

Expenditure.

	£	s.	d.	£	s.	d.
By Administration Expenditure				2,487	3	5
,, Salaries, Wages, Allowances, and Insurances of Staff (including Director)	29,125	5	0			
,, Annual Transfer to Pension Fund	525	0	0			
,, Advisory Committee on Aeronautics — Honoraria and Travelling Expenses	974	1	7			
				30,624	6	7
,, General Expenses				2,323	12	0
,, Physics Department, Research and Test Expenses				899	5	4
,, Metallurgical Department, Research and Test Expenses				589	9	8
,, Engineering Department— Research and Test Expenses				274	5	9
,, William Froude National Tank— Miscellaneous Expenditure				120	13	1
,, Road Board— Miscellaneous Expenses				50	12	6
,, Purchase of Instruments, &c., for Colonial and Foreign Institutions on Commission				91	7	4
Insurance of Instruments, &c., sent for Test ...				201	0	5
,, Tide Prediction Expenses				27	6	11
,, Meteorological and Hydroaeroplane Research in relation to Aeronautics, April, 1914, to April, 1915				500	0	0
,, Aeronautics—Additional Investigation				476	13	1
,, Transfers to Equipment Fund for Maintenance and Additions				756	2	6
,, Transfer to Building Fund...				100	0	0
,, Reserve for Bad and Doubtful Debts				50	13	3
,, Balance, being Excess of Income over Expenditure				80	13	8
				£39,653	5	6

THE NATIONAL PHYSICAL LABORATORY.

BALANCE SHEET, MARCH 31, 1915.

Liabilities.

	£	s.	d.	£	s.	d.
To **Capital Accounts**—						
Total Receipts, as per separate Accounts:—						
Building Fund	91,095	13	8			
Equipment Fund	20,289	10	8			
Aeronautics Building and Equipment Fund ...	17,936	1	2			
William Froude National Tank Building Fund ...	23,034	1	2*			
Road Board Laboratory Building and Equipment Fund	3,494	19	1			
				155,850	5	9
" **Pension Fund**				8,248	14	0

Assets.

	£	s.	d.	£	s.	d.
By **Capital Accounts**—						
Capital Expenditure as per separate accounts:—						
Building Fund	91,083	11	5			
Equipment Fund	18,758	13	2			
Aeronautics Building and Equipment Fund ...	17,189	11	6			
William Froude National Tank Building Fund ...	23,034	1	2			
Road Board Laboratory Building and Equipment Fund	3,432	5	4			
				153,498	2	7
Cash Balances at Bank of England	2,351	18	10			
Ditto London County and Westminster Bank, Ltd. ...	0	4	4			
				2,352	3	2
				155,850	5	9
" **Pension Fund**—Investments ...	8,194	3	8			
Ditto Cash Balance ...	54	10	4			
				8,248	14	0

	£	s.	d.	£	s.	d.
,, **Income Accounts**—						
Sundry Creditors	1,517	2	3			
Suspense Accounts	2,067	11	0			
				3,584	13	3
,, **Balance,** being Excess of Assets over Liabilities on Income Accounts...				2,347	5	3
				£170,030	18	3

	£	s.	d.	£	s.	d.	£	s.	d.
,, **Income Accounts**—									
Sundry Debtors—									
Tests				2,050	6	5			
Commissions				13	3	5			
Special Investigations and other work				269	5	0			
Sale of Publications... ...				17	16	1			
Insurance Claims				18	12	8			
Insurance, &c., paid in advance				440	5	0			
Proportion of Subscriptions falling due and not yet received from Tank Guarantors...				69	3	9			
Redpath, Brown & Co., Ltd., Expenditure to date on account of Concrete Testing Machine				90	15	0			
				2,969	7	4			
Less Reserve for Discounts...	50	17	2						
,, Reserve for Bad and Doubtful Debts ...	89	9	2						
				140	6	4			
							2,829	1	0
Stock of Chandlery Sundries							19	10	11
Cash at Bank and in hand							3,083	6	7
							£170,030	18	3

* Of this amount £1,660 14*s.* 4*d.* is a loan free of interest.

The following Table shows the progress and state of the Society with respect to the number of Fellows as at October 30, 1915:—

	Patron and Royal.	Foreign.	Compounders.	£4 yearly.	£3 yearly.	Total.
Oct. 31, 1914 ..	3	42	78	49	346	515
Since Elected ..			+ 2	+ 1	+ 14	+ 17
Since Deceased ..		− 4	− 5	− 3	− 9	− 21
Total ..	3	38	75	47	351	511

Account of Grants from the Donation Fund in 1914-15:—

	£	s.	d.
In aid of the expenses of moving the astronomical observatory of the Rev. T. R. Phillips...	25	0	0
In aid of the expenses of the International Catalogue of Scientific Literature	1,100	0	0
	£1,125	0	0

MINUTES OF MEETINGS OF THE ROYAL SOCIETY.

January 21, 1915.

Sir WILLIAM CROOKES, O.M., D.Sc., President, in the Chair.

In pursuance of the Statutes, the names of Candidates for election into the Society were read as follows:—

Andrewes, Frederick William.
Appleyard, Rollo.
Arber, Edward Alexander Newell.
Armstrong, Edward Frankland.
Ashworth, James Hartley.
Bainbridge, Francis Arthur.
Bairstow, Leonard.
Barton, Edwin Henry.
Bedson, Peter Phillips.
Bolton, Charles.
Bose, Jagadis Chunder.
Bousfield, William Robert.
Broom, Robert.
Brown, Sidney George.
Bryce, Thomas Hastie.
Buckmaster, George Alfred.
Budge, Ernest A. Wallis.
Carpenter, Henry Cort Harold.
Carslaw, Horatio Scott.
Coker, Ernest George.
Conway, Arthur William.
Conyngham, Gerald Ponsonby Lenox.
Dakin, Henry Drysdale.
Davidson, Sir James Mackenzie.
Dawson, Charles.
Dixon, Augustus Edward.
Dobell, Cecil Clifford.
Doncaster, Leonard.
Duckworth, Wynfrid Laurence Henry.
Edkins, John Sydney.
Evershed, John.
Ewart, Alfred James.
Fletcher, Walter Morley.
Forbes, Henry Ogg.
Fowler, George Herbert.
Gardner, John Addyman.
Gray, Albert Alexander.
Green, Arthur George.
Groom, Percy.
Guppy, Henry Brougham.
Halm, Jacob Karl Ernst.
Harmer, Frederic William.
Hartog, Marcus.
Hayden, Henry Hubert.
Henderson, George Gerald.
Henry, Augustine.
Hewlett, Richard Tanner.
Horrocks, William Heaton.
Houston, Alexander Cruikshank.
Hoyle, William Evans.
Hutchinson, Arthur.
Jenkinson, John Wilfrid.
Jordan, Karl.
Kennedy, Robert.

Kitchin, Finlay Lorimer.
Lake, Philip.
Laurie, Arthur Pillans.
Lewes, Vivian B.
Lewis, Thomas.
Locke, Frank Spiller.
Macdonald, John Smith.
McFadyean, Sir John.
McKenzie, Alexander.
Mackenzie, James.
McLennan, John Cunningham.
MacWilliam, John Alexander.
Maiden, Joseph Henry.
Marshall, Francis Hugh Adam.
Masterman, Arthur Thomas.
Matthews, Sir Thomas.
Maw, William Henry.
Morgan, Gilbert Thomas.
Murphy, Sir Shirley Forster.
Myers, Charles Samuel.
Newsholme, Arthur.
Nicholson, John William.
Ogilvie-Grant, William Robert.
Orton, Kennedy Joseph Previté.
Parsons, Hon. Richard Clere.
Pearson, Henry Harold Welch.
Peddie, William.
Pembrey, Marcus Seymour.
Perkins, Robert Cyril Layton.
Pickard, Robert Howson.
Pollock, James Arthur.
Purey-Cust, Herbert Edward.
Rây, Prafulla Chandra.
Read, Sir Charles Hercules.
Redwood, Sir Boverton.
Reed, Frederick Richard Cowper.
Reynolds, Sidney Hugh.
Rogers, Sir Leonard.
Rose, Sir Thomas Kirke.
Rosenheim, Otto.
Russell, Alexander.
Seligmann, Charles Gabriel.
Shattock, Samuel George.
Shearer, Cresswell.
Sidgwick, Nevil Vincent.
Simpson, George Clarke.
Smiles, Samuel.
Smith, Frank Edward.
Smith, Geoffrey Watkin.
Steele, Bertram Dillon.
Stephens, John William Watson.
Swinton, Alan Archibald Campbell.
Tansley, Arthur George.
Vaughan, Arthur.
Vaughan, David T. Gwynne.
Walker, Ernest William Ainley.
Walker, James.
Warren, William Henry.
Weizmann, Charles.
Wheeler, Richard Vernon.
Whytlaw-Gray, Robert.
Willis, John Christopher.
Wilson, Ernest.
Woods, Henry.
Yarrow, Alfred Fernandez.

The following Papers were read:—

I. "Atmospheric Electricity Potential Gradient at Kew Observatory, 1898-1912." By Dr. C. CHREE, F.R.S.

II. "The Transmission of Electric Waves over the Surface of the Earth." By Prof. A. E. H. LOVE, F.R.S.

III. "Electromagnetic Waves in a Perfectly Conducting Tube." By L. SILBERSTEIN. Communicated by Prof. A. W. PORTER, F.R.S.

IV. "An Electrically Heated Full Radiator." By H. B. KEENE. Communicated by Sir O. LODGE, F.R.S.

January 28, 1915.

Sir WILLIAM CROOKES, O.M., D.Sc., President, in the Chair.

The following Papers were read:—

I. "The Influence of Salt-concentration on Hæmolysis." By W. W. C. TOPLEY. Communicated by Prof. F. W. MOTT, F.R.S.

II. "The Life-cycle of *Cladocera*, with Remarks on the Physiology of Growth and Reproduction in Crustacea." By G. SMITH. Communicated by E. S. GOODRICH, F.R.S.

III. "Investigations on Protozoa in Relation to the Factor limiting Bacterial Activity in Soil." By T. GOODEY. Communicated by Prof. F. W. GAMBLE, F.R.S.

IV. "On the Mesodermic Origin and the Fate of the so-called Mesectoderm in Petromyzon." By S. HATTA. Communicated by Prof. E. W. MACBRIDE, F.R.S.

V. "The Influence of Homodromous and Heterodromous Electric Currents on Transmission of Excitation in Plant and Animal." By Prof. J. C. BOSE. Communicated by Prof. S. H. VINES, F.R.S.

February 4, 1915.

Sir WILLIAM CROOKES, O.M., D.Sc., President, in the Chair.

The following Papers were read:—

I. "Discontinuous Fluid Motion past a Bent Plane, with Special Reference to Aeroplane Problems." By Prof. G. H. BRYAN, F.R.S., and R. JONES.

II. "A New Type of Series in the Band Spectrum associated with Helium." By Prof. A. FOWLER, F.R.S.

III. "On the Spectra of Ordinary Lead and Lead of Radioactive Origin." By T. R. MERTON. Communicated by Prof. A. FOWLER, F.R.S.

IV. "On the Viscosity of the Vapour of Iodine." By A. O. RANKINE. Communicated by Prof. A. W. PORTER, F.R.S.

February 11, 1915.

Sir WILLIAM CROOKES, O.M., D.Sc., President, in the Chair.

The following Papers were read :—

I. "*Lepidostrobus Kentuckiensis*, formerly *Lepidostrobus Fischeri*, Scott and Jeffrey.—A Correction." By Dr. D. H. SCOTT, For. Sec. R.S.

II. "The Excitatory Process in the Dog's Heart. Part II.—The Ventricles." By T. LEWIS and M. A. ROTHSCHILD. Communicated by Prof. E. H. STARLING, F.R.S.

III. "On the Variation in the Growth of Mammalian Tissue *in vitro* according to the Age of the Animal." By A. J. WALTON. Communicated by Prof. W. BULLOCH, F.R.S.

February 18, 1915.

Sir WILLIAM CROOKES, O.M., D.Sc., President, in the Chair.

The following Papers were read :—

I. "Gaseous Combustion at High Pressures." By Prof. W. A. BONE, F.R.S., and others.

II. "The Orbits of a Charged Particle round an Electric and Magnetic Nucleus." By Prof. W. M. HICKS, F.R.S.

III. "The Lunar Diurnal Magnetic Variation and its Change with Lunar Distance." By S. CHAPMAN. Communicated by Sir F. W. DYSON, F.R.S.

IV. "Some Temperature Refraction Coefficients of Optical Glass." By Lt.-Col. J. W. GIFFORD. Communicated by Prof. S. P. THOMPSON, F.R.S.

February 25, 1915.

Sir WILLIAM CROOKES, O.M., D.Sc., President, in the Chair.

The following Papers were read:—

I. "The Effect of the Depth of Pulmonary Ventilation on the Oxygen in the Venous Blood of Man." By Prof. L. HILL, F.R.S., and J. F. TWORT.

II. "The Effect of Functional Activity upon the Metabolism, Blood Flow and Exudation in Organs." By J. BARCROFT, F.R.S. and TOYOJIRO KATO.

III. "The Osmotic Balance of Skeletal Muscle." By Miss D. JORDAN LLOYD. Communicated by W. B. HARDY, F.R.S.

IV. "On the Function of Chlorophyll." By A. J. EWART, D.Sc. Communicated by Prof. A. C. SEWARD, F.R.S.

V. "The Influence of the Hydrogen Concentration upon the Optimum Temperature of a Ferment." By A. COMPTON. Communicated by Prof. W. BULLOCH, F.R.S.

VI. "Functional Œdema in Frog's Muscle." By M. BACK, K. M. COGAN, and A. E. TOWERS. Communicated by J. BARCROFT, F.R.S.

March 4, 1915.

Sir WILLIAM CROOKES, O.M., D.Sc., President, in the Chair.

In pursuance of the Statutes, the names of the Candidates recommended for election into the Society were read as follows:—

Andrewes, Frederick William.
Conway, Arthur William.
Doncaster, Leonard.
Evershed, John.
Fletcher, Walter Morley.
Green, Arthur George.
Hayden, Henry Hubert.
Mackenzie, James.
McLennan, John Cunningham.
Masterman, Arthur Thomas.
Morgan, Gilbert Thomas.
Myers, Charles Samuel.
Simpson, George Clarke.
Swinton, Alan Archibald Campbell.
Tansley, Arthur George.

The following Papers were read:—

I. "A Bolometric Method of Determining the Efficiency of Radiating Bodies." By Prof. W. A. BONE, F.R.S., Prof. H. L. CALLENDAR, F.R.S., and H. J. YATES.

II. "The Simplification of the Arithmetical Processes of Involution and Evolution." By E. CHAPPELL. Communicated by Prof. W. E. DALBY, F.R.S.

III. "The Elastic Properties of Steel at Moderately High Temperatures." By F. E. ROWETT. Communicated by Prof. B. HOPKINSON, F.R.S.

IV. "The Laws of Series Spectra." By Prof. J. W. NICHOLSON. Communicated by Prof. A. FOWLER, F.R.S.

March 11, 1915.

Sir WILLIAM CROOKES, O.M., D.Sc., President, followed by Mr. W. B. HARDY, Vice-President, in the Chair.

The following Papers were read:—

I. "Contributions to the Study of the Bionomics and Reproductive Processes of the Foraminifera." By E. HERON-ALLEN. Communicated by Prof. W. A. HERDMAN, F.R.S.

II. "On the Occurrence of an Intracranial Ganglion upon the Oculomotor Nerve in *Scyllium canicula*, with a Suggestion as to its Bearing upon the Question of the Segmental Value of Certain of the Cranial Nerves." By G. E. NICHOLLS. Communicated by Prof. A. DENDY, F.R.S.

III. "Experiments on the Restoration of Paralysed Muscles by Means of Nerve Anastomosis. Part III.—Anastomosis of the Brachial Plexus, with a Consideration of the Distribution of its Roots." By Prof. R. KENNEDY. Communicated by Prof. J. G. McKENDRICK, F.R.S.

IV. "On the Mechanism of the Cardiac Valves.—A Preliminary Communication." By A. F. S. KENT. Communicated by Prof. C. S. SHERRINGTON, F.R.S.

March 18, 1915.

Sir WILLIAM CROOKES, O.M., D.Sc., President, in the Chair.

The Bakerian Lecture was delivered by Prof. W. H. BRAGG, F.R.S., on "X-rays and Crystals."

March 25, 1915.

Sir WILLIAM CROOKES, O.M., D.Sc., President, in the Chair.

The following Papers were read:—

I. "The Production of Growths or Deposits in Metastable Inorganic Hydrosols." By Prof. B. MOORE, F.R.S.

II. "On Forms of Growth resembling Living Organisms and their Products slowly deposited from Metastable Solutions of Inorganic Colloids." By Prof. B. MOORE, F.R.S., and W. G. EVANS.

III. "A Contribution to our Knowledge of the Chemistry of Coat-colour in Animals and of Dominant and Recessive Whiteness." By the Hon. H. ONSLOW. Communicated by Dr. F. G. HOPKINS, F.R.S.

April 22, 1915.

Sir WILLIAM CROOKES, O.M., D.Sc., President, in the Chair.

The following Papers were read:—

I. "Deep Water Waves, Progressive or Stationary, to the Third Order of Approximation." By Lord RAYLEIGH, O.M., F.R.S.

II. "A Chemically Active Modification of Nitrogen, produced by the Electric Discharge.—VI." By the Hon. R. J. STRUTT, F.R.S.

III. "The Difference between the Magnetic Diurnal Variations on Ordinary and Quiet Days at Kew Observatory." By Dr. C. CHREE, F.R.S.

IV. "The Effects of Different Gases on the Electron Emission from Glowing Solids." By F. HORTON. Communicated by Sir J. J. THOMSON, F.R.S.

V. "Heats of Dilution of Concentrated Solutions." By W. S. TUCKER. Communicated by Prof. H. L. CALLENDAR, F.R.S.

VI. "On the Origin of the '4686' Series." By T. R. MERTON. Communicated by Prof. A. FOWLER, F.R.S.

April 29, 1915.

Sir WILLIAM CROOKES, O.M., D.Sc., President, in the Chair

The following Papers were read:—

I. "The Transmission of Infra-red Rays by the Media of the Eye, the Transmission of Radiant Energy by Crookes and other Glasses." By H. HARTRIDGE and A. V. HILL. Communicated by W. B. HARDY, F.R.S.

II. "Surface Tension and Ferment Action." By E. BEARD and W. CRAMER. Communicated by Sir E. SCHÄFER, F.R.S.

III. "Surface Tension as a Factor controlling Cell Metabolism." By W. CRAMER. Communicated by Sir E. SCHÄFER, F.R.S.

May 6, 1915.

Sir ALFRED KEMPE, Vice-President and Treasurer, in the Chair.

The following Papers were read:—

I. "Some Problems Illustrating the Forms of Nebulæ." By G. W. WALKER, F.R.S.

II. "Observations on the Resonance Radiation of Sodium Vapour." By the Hon. R. J. STRUTT, F.R.S.

III. "Local Differences of Pressure near an Obstacle in Oscillating Water." By HERTHA AYRTON. Communicated by Lord RAYLEIGH, O.M., F.R.S.

May 13, 1915.

Annual Meeting for the Election of Fellows.

Sir WILLIAM CROOKES, O.M., D.Sc., President, in the Chair.

The Statutes relating to the Election of Fellows having been read, Mr. W. Barlow and Prof. H. F. Newall were, with the consent of the Society, nominated Scrutators, to assist the Secretaries in the examination of the balloting lists.

The votes of the Fellows present were collected, and the following Candidates were declared duly elected into the Society:—

Andrewes, Frederick William.
Conway, Arthur William.
Doncaster, Leonard.
Evershed, John.
Fletcher, Walter Morley.
Green, Arthur George.
Hayden, Henry Hubert.
Mackenzie, James.
McLennan, John Cunningham.
Masterman, Arthur Thomas.
Morgan, Gilbert Thomas.
Myers, Charles Samuel.
Simpson, George Clarke.
Swinton, Alan Archibald Campbell.
Tansley, Arthur George.

May 13, 1915.

Sir WILLIAM CROOKES, O.M., D.Sc., President, in the Chair.

M. H. L. Le Chatelier, Foreign Member (elected 1913), was admitted into the Society.

The following Papers were read:—

I. "The Development of the Thymus, Epithelial Bodies, and Thyroid in the Vulpine Phalanger (*Trichosurus vulpecula*)." By ELIZABETH A. FRASER and Prof. J. P. HILL, F.R.S.

II. "Some Observations on the Development of the Thymus, Epithelial Bodies, and Thyroid in Phascolarctos, Phascolomys, and Perameles." By ELIZABETH A. FRASER. Communicated by Prof. J. P. HILL, F.R.S.

III. "Measurement of the Specific Heat of Steam at Atmospheric Pressure and 104·5° C." By J. H. BRINKWORTH. Communicated by Prof. H. L. CALLENDAR, F.R.S.

IV. "Thermal Properties of Carbonic Acid at Low Temperatures.—II." By C. F. JENKIN and D. R. PYE. Communicated by Sir A. EWING, F.R.S.

May 20, 1915.

Sir WILLIAM CROOKES, O.M., D.Sc., President, in the Chair.

Prof. F. W. Andrewes, Mr. L. Doncaster, Dr. W. M. Fletcher, Dr. James Mackenzie, and Mr. A. G. Tansley were admitted into the Society.

The following Papers were read :—

I. "On the Corpuscular Radiation Liberated in Vapours by Homogeneous X-radiation." By H. Moore. Communicated by Prof. O. W. Richardson, F.R.S.

II. "The Absorption in Lead of γ-Rays emitted by Radium B and Radium C." By H. Richardson. Communicated by Sir E. Rutherford, F.R.S.

III. "On the Application of Interference Methods to the Study of the Origin of Certain Spectrum Lines." By T. R. Merton. Communicated by Prof. A. Fowler, F.R.S.

The Society adjourned over the Whitsun Recess to Thursday, June 3.

June 3, 1915.

Sir WILLIAM CROOKES, O.M., D.Sc., President, in the Chair.

Prof. G. T. Morgan and Mr. Campbell Swinton were admitted into the Society.

The following Papers were read :—

I. "On the Shapes of the Equipotential Surfaces in the Air near Long Walls or Buildings and on their Effect on the Measurement of Atmospheric Potential Gradients." By Prof. C. H. Lees, F.R.S.

II. "The Influence of Gases on the Emission of Electrons and Ions from Hot Metals." By Prof. O. W. Richardson, F.R.S.

III. "The Band Spectrum associated with Helium." By J. W. Nicholson. Communicated by Prof. A. Fowler, F.R.S.

IV. "The Chromosome Cycle in Coccidia and Gregarines." By C. Dobell and A. P. Jameson. Communicated by Prof. J. B. Farmer, F.R.S.

The Society adjourned to June 17.

June 17, 1915.

Sir WILLIAM CROOKES, O.M., D.Sc., President, in the Chair.

Prof. A. W. Conway, Prof. A. G. Green, and Dr. A. T. Masterman were admitted into the Society.

The following Papers were read:—

I. "Soil Protozoa and Soil Bacteria." By E. J. RUSSELL. Communicated by H. T. BROWN, F.R.S.

II. "On the Enhanced Series of Lines in Spectra of the Alkaline Earths." By Prof. W. M. HICKS, F.R.S.

III. "On Certain Linear Differential Equations of Astronomical Interest." By Prof. H. F. BAKER, F.R.S.

IV. "On the Partial Correlation-Ratio." By Prof. KARL PEARSON, F.R.S.

V. "The Effect of Temperature on the Hissing of Water when Flowing through a Constricted Tube." By S. SKINNER and F. ENTWISTLE. Communicated by Sir NAPIER SHAW, F.R.S.

VI. "Ionisation Potentials of Mercury, Cadmium, and Zinc, and the Single and Many-lined Spectra of these Elements." By J. C. McLENNAN, F.R.S., and J. P. HENDERSON.

VII. "The Monoclinic Sulphates containing Ammonium.—Completion of the Double Sulphate Series." By A. E. H. TUTTON, F.R.S.

VIII. "General Equations for the Neutralisation of Dibasic Acids, and their Use to Calculate the Acidity of Dilute Carbonate Solutions." By E. B. R. PRIDEAUX. Communicated by Prof. F. G. DONNAN, F.R.S.

IX. "The Electrical Conductivity and Luminosity of Flames containing Salt Vapours." By Prof. H. A. WILSON, F.R.S.

X. "On a Spectrum associated with Carbon in Relation to the Wolf-Rayet Stars." By T. R. MERTON. Communicated by Prof. A. FOWLER, F.R.S.

XI. "The Threshold of Vision for Different Coloured Lights." By Sir WM. ABNEY, F.R.S., and Prof. W. WATSON, F.R.S.

XII. "Hydrodynamical Problems suggested by Pitot's Tubes." By LORD RAYLEIGH, O.M., F.R.S.

XIII. "Electrical Effects accompanying the Decomposition of Organic Compounds. II.—Ionisation of the Gases produced during Fermentation." By M. C. POTTER. Communicated by Dr. A. D. WALLER, F.R.S.

XIV. "The Inheritance of Colour in the Stick-Insect (*Carausius morosus*)." By Prof. E. W. MACBRIDE, F.R.S., and A. JACKSON.

XV. "On the Relation between Transpiration and Stomatal Aperture." By Sir FRANCIS DARWIN, F.R.S.

XVI. "The Monotreme Skull—a Contribution to Mammalian Morphogenesis." By D. M. S. WATSON. Communicated by Prof. J. P. HILL, F.R.S.

The Society adjourned over the Long Vacation to Thursday, November 4.

MINUTES OF MEETINGS.—SESSION 1915-16.

November 4, 1915.

Sir WILLIAM CROOKES, O.M., D.Sc., President, in the Chair.

The following Papers were read:—

I. "A Diagram to Facilitate the Study of External Ballistics." By Prof. W. E. DALBY, F.R.S.

II. "An Application of the Principle of Dynamical Similitude to Molecular Physics." By W. B. HARDY, F.R.S.

III. "The Motion of a Stream of Finite Depth past a Body." By R. JONES. Communicated by Dr. R. T. GLAZEBROOK, F.R.S.

IV. "On Deep-Sea Water Waves Caused by a Local Disturbance on or beneath the Surface." By K. TERAZAWA. Communicated by Sir J. LARMOR, F.R.S.

V. "The Consumption of Carbon in the Electric Arc." By W. G. DUFFIELD. Communicated by Sir E. RUTHERFORD, F.R.S.

The following Papers, published during the vacation, were read in title:—

I. "Observations on the Fluorescence and Resonance of Sodium Vapour.—II." By the Hon. R. J. STRUTT, F.R.S.

II. "Some Applications of Conformal Transformation to Problems in Hydrodynamics. Supplementary Note." By J. G. LEATHEM. Communicated by Sir J. LARMOR, F.R.S.

III. "On the Conditions under which the 'Probable Errors' of Frequency Distributions have a Real Significance." By L. ISSERLIS. Communicated by Prof. KARL PEARSON, F.R.S.

IV. "The Reaction between Gas and Pole in the Electrical Ignition of Gaseous Mixtures." By Prof. W. M. THORNTON. Communicated by Sir C. PARSONS, F.R.S.

V. "On the Longitudinal Strength of Cylinders closed by Screw Plugs." By Lieut.-Colonel A. G. HADCOCK. Communicated by Sir G. GREENHILL, F.R.S.

VI. "Some Applications of Conformal Transformation to Problems in Hydrodynamics." By J. G. LEATHEM. Communicated by Sir J. LARMOR, F.R.S.

VII. "Volatilisation of Extremely Thin Radioactive Deposits." By A. B. WOOD. Communicated by Sir E. RUTHERFORD, F.R.S.

VIII. "Some Experiments on the Properties of Dielectrics." By S. W. RICHARDSON. Communicated by Sir J. J. THOMSON, F.R.S.

IX. "A Bacterial Test for Plant-Food Accessories (Auximones)." By W. B. BOTTOMLEY. Communicated by Prof. F. W. OLIVER, F.R.S.

November 11, 1915.

Sir WILLIAM CROOKES, O.M., D.Sc., President, in the Chair.

In pursuance of the Statutes, notice of the ensuing Anniversary Meeting was given from the Chair.

The following Papers were read :—

I. "Effects of Functional Activity in Striated Muscle and the Submaxillary Gland." By J. BARCROFT, F.R.S., and T. KATO.

II. "Analyses of Agricultural Yield. Part II.—The Sowing-date Experiment with Egyptian Cotton, 1913." By W. L. BALLS and F. S. HOLTON. Communicated by Dr. F. S. BLACKMAN, F.R.S.

III. "On Williamsoniella, a New Type of Bennettitalean Flower." By H. H. THOMAS. Communicated by Prof. A. C. SEWARD, F.R.S.

IV. "Studies on *a priori* Pathometry.—Part I." By Sir RONALD ROSS, F.R.S.

V. "The Spread of the Excitatory Process in the Vertebrate Heart. Part I.—The Toad's Ventricle. Part II.—The Tortoise Ventricle. Part III.—The Dog's Ventricle. Part IV.—The Human Ventricle. Part V.—The Bird's Heart." By T. LEWIS. Communicated by Prof. E. H. STARLING, F.R.S.

November 18, 1915.

Sir WILLIAM CROOKES, O.M., D.Sc., President, in the Chair.

In pursuance of the Statutes, Sir A. Pedler, Dr. A. Strahan, and Prof. J. M. Thomson were elected Auditors of the Treasurer's accounts on the part of the Society.

In pursuance of the Statutes, notice of the ensuing Anniversary Meeting was given from the Chair.

The following Papers were read:—

I. "On the Theory of the Capillary Tube." By Lord RAYLEIGH, O.M., F.R.S.

II. "On the Effect of the Form of the Transverse Section on the Resistance to the Motion of an Elongated Body Parallel to its Length through a Fluid whose Viscosity is not Negligible." By Prof. C. H. LEES, F.R.S.

III. "On a Method of Estimating Distances at Sea in Fog or Thick Weather." By Prof. J. JOLY, F.R.S.

IV. "On a Method of Avoiding Collision at Sea." By Prof. J. JOLY, F.R.S.

V. "The Flow of Electricity through Dielectrics." By S. W. RICHARDSON. Communicated by Prof. SCHUSTER, Sec. R.S.

VI. "On the Kinetic Theory of Gaseous Viscosity and Thermal Conduction, and the Law of Distribution of Molecular Velocities in the Disturbed State." By S. CHAPMAN. Communicated by Sir J. LARMOR, F.R.S.

November 25, 1915.

Sir WILLIAM CROOKES, O.M., D.Sc., President, in the Chair.

A Special General Meeting was held to discuss the Annual Report of the Council.

In pursuance of the Statutes, notice of the ensuing Anniversary Meeting was given from the Chair, and the list of Officers and Council nominated for election was read as follows :—

President.—Sir Joseph J. Thomson, O.M., M.A., D.Sc., LL.D.

Treasurer.—Sir Alfred Bray Kempe, M.A., D.C.L.

Secretaries.—{ Prof. Arthur Schuster, Sc.D., Ph.D.
William Bate Hardy, M.A.

Foreign Secretary.—Dukinfield Henry Scott, M.A., Ph.D., LL.D.

Other Members of the Council.—Prof. John George Adami, M.D.; Sir Thomas Clifford Allbutt, K.C.B.; Frederick Frost Blackman, D.Sc.; Dugald Clerk, D.Sc.; Sir William Crookes, O.M., D.Sc.; Prof. Arthur Dendy, D.Sc.; Henry John Horstman Fenton, Sc.D.; John Stanley Gardiner, M.A.; Henry Head, M.D.; George William Lamplugh; Prof. Augustus Edward Hough Love, D.Sc.; Major Percy Alexander MacMahon, D.Sc.; Prof. Arthur Smithells, B.Sc.; Prof. Ernest Henry Starling, M.D.; Richard Threlfall, M.A.; Sir Philip Watts, K.C.B.

The following Papers were read:—

I. "The Measurement of the Rate of Heat Loss at Body Temperature by Convection, Radiation, and Evaporation." By M. FLACK, O. W. GRIFFITH, and L. HILL, F.R.S.

II. "The Growth of the Body in Man.—The Relationship between the Body Weight and the Body Length (Stem Length)." By E. W. A. WALKER. Communicated by Prof. C. S. SHERRINGTON, F.R.S.

III. "The Rate of Absorption of various Phenolic Solutions by Seeds of *Hordeum vulgare* and the Factors Governing the Rate of Diffusion of Aqueous Solutions across Semipermeable Membranes." By Prof. A. J. BROWN, F.R.S., and F. TINKER.

IV. "The Controlling Influence of Carbon Dioxide. Part III.—The Retarding Effect of Carbon Dioxide on Respiration." By F. KIDD. Communicated by Dr. F. F. BLACKMAN, F.R.S.

November 30, 1915.

ANNIVERSARY MEETING.

Sir WILLIAM CROOKES, O.M., D.Sc., President, in the Chair.

The Report of the Auditors of the Treasurer's accounts was read, and the thanks of the Society were given to the Treasurer and to the Auditors.

The List of Fellows deceased and the List of Fellows elected into the Society since the last Anniversary were read.

The Report to the Society from the Council, upon the work during the past year, was, upon the motion of the President, received.

The President delivered his Anniversary Address. On the motion of Sir Archibald Geikie, seconded by Prof. Emerson Reynolds, the thanks of the Society were voted to the President for his Address, and he was requested to allow it to be printed.

The Awards of the Medals for the year were announced as follows, and the Medals were presented from the Chair :—

The Copley Medal.........	To Prof. Ivan P. Pavlov, For. Mem. R.S.
A Royal Medal	,, Sir Joseph Larmor, F.R.S.
A Royal Medal	,, Dr. W. H. R. Rivers, F.R.S.
The Davy Medal	,, Prof. Paul Sabatier.
The Hughes Medal	,, Prof. Paul Langevin.

The President having, with the consent of the Society, nominated Dr. Teall and Prof. H. H. Turner as Scrutators, to assist the Secretaries in examining the balloting lists for the election of Council and Officers, the votes of the Fellows present were taken. The Scrutators reported that the Council and Officers nominated at the preceding meeting had been duly elected, and their names were accordingly announced from the Chair.

The following Fellows were present and voted :—

J. G. Adami.
Col. A. W. Alcock.
C. W. Andrews.
W. Barlow.
W. M. Bayliss.
F. E. Beddard.
H. Brown.
Sir D. Bruce.
Sir W. H. M. Christie.
Capt. E. W. Creak.
Sir W. Crookes.
A. R. Cushny.
C. Cuthbertson.
A. Dendy.
Sir F. W. Dyson.
Sir D. Ferrier.

A. Fowler.
J. S. Gamble.
E. J. Garwood.
Sir A. Geikie.
A. G. Green.
W. B. Hardy.
J. A. Harker.
H. S. Hele-Shaw.
Major E. H. Hills.
G. J. Hinde.
J. H. Jeans.
Sir A. B. Kempe.
G. W. Lamplugh.
Sir R. Lankester.
Sir J. Larmor.
A. E. H. Love.
Capt. H. G. Lyons.
E. W. MacBride.
E. T. Newton.
R. D. Oldham.
Sir A. Pedler.
J. Perry.
J. Emerson Reynolds.
W. H. R. Rivers.
A. Schuster.
A. Scott.
D. H. Scott.
W. J. S. Smith.
A. Strahan.
A. A. C. Swinton.
J. J. H. Teall.
S. P. Thompson.
J. Thorpe.
Sir W. Tilden.
H. H. Turner.
A. D. Waller.
W. W. Watts.
A. N. Whitehead.
C. T. R. Wilson.
A. Smith Woodward.
H. Woodward.
Sir J. Wolfe-Barry.

The thanks of the Society were given to the Scrutators.

December 2, 1915.

Sir J. J. THOMSON, O.M., President, in the Chair.

The following Papers were read :—

I. "Note on the Existence of Converging Sequences in Certain Oscillating Successions of Functions." By W. H. YOUNG, F.R.S.

II. "On the Emulsifying Action of Soap: A Contribution to the Theory of Detergent Action." By S. A. SHORTER and E. ELLINGWORTH. Communicated by Prof. W. H. BRAGG, F.R.S.

III. "The Newtonian Constant of Gravitation as affected by Temperature." By P. E. SHAW. Communicated by C. V. BOYS, F.R.S.

IV. "Skin Friction of the Wind on the Earth's Surface." By G. I. TAYLOR. Communicated by Sir N. SHAW, F.R.S.

December 9, 1915.

Sir J. J. THOMSON, O.M., President, in the Chair.

The President announced that he had appointed as Vice-Presidents :—

Sir Alfred Kempe (Treasurer),
Sir Clifford Allbutt,
Sir William Crookes,
Sir Philip Watts.

The Croonian Lecture for 1915, on "The Respiratory Process in Muscle, and the Nature of Muscular Motion," was delivered by Dr. W. M. FLETCHER, F.R.S., and Prof. F. G. HOPKINS, F.R.S.

The Society adjourned to Thursday, January 27, 1916.

INDEX.

Made in United States
Troutdale, OR
11/03/2026

56844500R00136